HVAC

MASTERY BIBLE

An All-in-One Guide to Heating, Ventilation, and Air Conditioning
From Installation to Repair
Integrate Smart Home Technologies
Understand Tax Credits Advantages

Gregory T. Watts

TABLE OF CONTENTS

Introduction

THE WORLD OF HVAC

Ladies and gentlemen, DIY aficionados, and masters of craftsmanship, as we gather here today, let's set the stage for a topic that touches the very core of our existence - the air we breathe. Yes, that invisible, omnipresent force which, much like the finest details in your most cherished DIY project, plays an instrumental role in the backdrop of our lives.

While our hands are busy sculpting, painting, and perfecting, our lungs are tirelessly at work, drawing in the essence of life. But here's a thought to ponder upon: As a passionate DIYer, how often do we pause to consider the quality of the very air we take in? Think of your body as a finely crafted masterpiece, a work of art that's continuously being shaped. What would happen if you supplied this artwork with compromised materials? Just as a smudged blueprint can blur your project's vision, subpar air quality can fog our mental faculties and impair our health. It's time we recognize the undeniable truth - pristine air quality is not a luxury but a necessity. Clear air equals clear health, a fundamental principle that cannot be overlooked.

Breathing is our most vital process, and its quality influences not just our physical health but also our mental resilience. Every inhalation of polluted air is akin to a small error in a project, minor at first glance but with potential long-term consequences. It can dull our mental toolkit, erode our motivation, and dampen the enthusiasm that fuels our DIY pursuits. The insidious pollutants sneaking into our environments aren't merely adversaries of our health; they wage war against our creativity, mood, and clarity. It's a long game, my friends. And in the intricate world of DIY, where attention to detail is paramount, we ought to prioritize the air we breathe with the same fervor we reserve for our projects.

Enter the unsung hero of modern living: the HVAC system. Amidst the myriad tools and gadgets that adorn our workspaces, HVAC systems silently labor away, ensuring our sanctuaries remain just that – havens of purity, comfort, and well-being. Envision air purification as the meticulous sanding of a piece – refining, perfecting, and elevating. The role of humidity, too, cannot be overlooked. Mastering it means mastering precision. After all, in the grand symphony of DIY, even a note out of place can disrupt the melody. Excessive humidity can be the ruin of your wooden masterpieces, while an environment too dry might rob them of their luster. The solution? Knowledge. The tools? Reliable humidifiers and dehumidifiers.

But wait, there's another unsung hero in our tale - ventilation. Often relegated to the background, adequate ventilation acts as the beating heart of any workspace, ensuring a continuous supply of fresh air and expelling the old. It stands guard, protecting us from the invisible, harmful fumes that might otherwise tarnish our DIY dreams.

Yet, HVAC's contributions aren't limited to health alone. Today, these systems are the epitome of personalized luxury and comfort. Their role in modern architectural marvels is undeniable. As the walls of our homes become canvases displaying our artistry, HVAC systems ensure these spaces remain inviting, cozy, and most importantly, healthy.

To truly master our domain, knowledge of HVAC is indispensable. Every modern home, every DIY workspace stands resiliently on the foundational knowledge of HVAC. And as we embark on this enlightening journey through the intricacies of heating, ventilation, and air conditioning, remember, it's not just about the creature comforts. It's about carving out future sanctuaries – spaces where every corner resonates with well-being, where every breath we draw is a testament to our commitment to excellence, precision, and the art of living.

In conclusion, while the realm of HVAC might seem vast and intricate, with your passion for DIY and thirst for knowledge, understanding its depths is well within reach. Let's walk this path together, ensuring that our homes, our sanctuaries, are not just shelters but true embodiments of health and comfort.

Thank you for embarking on this journey with us. Your quest for knowledge and precision in the DIY world is commendable, and we're honored to be a part of it. Here's to crafting a healthier, more comfortable tomorrow!

CHAPTER 1

WHY HVAC IS FUNDAMENTAL

Optimal Health and Well-being: The HVAC-Quality of Life Connection

The Symphony of Air Quality & Health: A Primer

Before diving into the nuts and bolts of HVAC, let's start with something fundamental: the air you breathe. Picture the air as an unseen fabric that intimately drapes over you, every second of every day. Clean, pure air acts like luxury silk – smooth, comforting, and utterly rejuvenating. It keeps your inner gears (lungs, heart, brain) running like a well-oiled DIY project. But, tainted air? That's akin to scratchy, low-grade wool, causing sneezes, coughs, and long-term wear on your body's machinery. The moral of our story? Quality air isn't a luxury; it's a non-negotiable for a healthy life. As we journey into the HVAC world, remember this: it's not just about gadgets and systems; it's about crafting a healthier living canvas, one breath at a time.

The Unsung Heroes: HVAC's Role in Sculpting Your Air

Let's talk about the unsung heroes quietly working in the background of our homes and offices: HVAC systems. Think of your indoor air as a blank canvas, and the HVAC as the artist, subtly shading in the details. From filtering out those pesky dust particles to artfully managing the moisture levels, HVAC systems mold the very essence of our indoor spaces. The heaters, coolers, and ventilators decide how thick or thin the air feels, while filters play gatekeepers, determining what gets in (or stays out). Ever noticed that fresh, crisp feeling after a system check-up? That's the HVAC, masterfully curating your air for optimal comfort and health. As we navigate through this guide, you'll gain a profound appreciation for this silent maestro, working tirelessly to perfect the composition of your air.

Understanding Airborne Pollutants

Unmasking the Invisible Invaders: Your Guide to Indoor Pollutants

As we dive deeper into the air we breathe, it's crucial we familiarize ourselves with the unseen guests in our spaces. Enter: indoor pollutants. These sneaky culprits range from microscopic dust mites dancing on our furniture to volatile organic compounds (VOCs) stealthily released from our fresh paint jobs or new furnishings. Don't forget about pollen, a seasonal visitor for many, or pet dander, the clingy remnants of our furry friends. And, lurking in the shadows might be mold spores, especially if humidity's been having a party. Understanding these common pollutants is the first step in ensuring our indoor atmospheres are as pristine as our well-crafted DIY projects. Buckle up, as we journey through the realm of what's floating amidst our indoor breezes!

Decoding the Origins: Where Do Our Indoor Pollutants Come From?

Ever wonder where those invisible airborne trespassers come from? Well, let's unravel this mystery. Our homes, as cozy as they are, can sometimes hide pollutants in plain sight. Building materials, for instance, can stealthily emit formaldehyde. That brand-new couch or drapery? They might be releasing VOCs, adding to the unseen cocktail in our rooms. Our trusty cleaning agents, while making surfaces gleam, might leave behind a trail of chemicals in the air. And, just when you thought the outdoors would be the culprit-free zone, in comes pollen—nature's own confetti—to join the mix. Knowledge is power, and by understanding these sources, we are a step closer to crafting cleaner, healthier spaces for our next DIY masterpiece.

Connecting the Dots: Pollutants and Their Stealthy Impact on Health

As we craft, build, and innovate, we sometimes forget the unseen challenges lurking around: indoor pollutants. These microscopic invaders can be real party crashers! From instigating sneezy allergies and aggravating asthma, they can even play long-con games, contributing to chronic respiratory conditions or heart ailments. It's like sanding wood – you might not notice the fine particles at first, but over time, they can cloud our environment and our health. And some of these pollutants don't just stop at physical health; they can impact our mood and cognition too. So, as we champion precision in our DIY endeavors, it's paramount to keep an eye (and nose!) out for these hidden health hijackers.

The Role of HVAC in Filtering and Reducing Pollutants

The Magic Behind the Scenes: HVAC's Air Filtration Wizardry

Imagine pouring a cup of tea and watching those tiny tea leaves staying behind, courtesy of your strainer. This is pretty much what our trusty HVAC systems do, but on a grander scale. These systems come equipped with specialized filters that act as sentinels, capturing dust, dander, and other airborne particles, ensuring we get nothing but crisp, clean air. The key is in the layers of fibrous or porous materials that trap and hold pollutants. It's like having a dedicated security guard for your lungs! For those who adore efficiency and attention to detail, understanding this marvel of HVAC design is both fascinating and essential. Cheers to cleaner breaths and clearer spaces!

Keep It Spick and Span: The Golden Rule for Peak Air Filtration

Alright DIY enthusiasts, picture this: you've got a high-performance car, but you never change its oil. Won't run smoothly for long, will it? Similarly, your HVAC's filtration prowess hinges on consistent upkeep. Regular maintenance isn't just about longevity; it's about ensuring every inhale is as pure as possible. Dust, pet dander, and other pollutants can clog filters over time, hampering efficiency. By cleaning or replacing filters and giving the system its deserved TLC, you're not just boosting performance but also safeguarding your health. After all, a well-tuned machine delivers exceptional results. Let's make 'cleanliness' our mantra for the freshest air in every DIY endeavor!

From Standard to Advanced: Unveiling the Magic of UV Light

There are HVAC upgrades that can elevate your system from "efficient" to "powerhouse"! We're talking about UV light, not just a nifty effect at blacklight parties, but a genuine pathogen-neutralizing hero. That's right. When correctly positioned within your HVAC setup, this ultraviolet light beam can obliterate bacteria, viruses, and molds, stopping them from freely circulating in the air you breathe. The science is riveting, but the core concept? It breaks down and disintegrates microscopic invaders. If you're looking to take your indoor environment up a notch in cleanliness and health, consider an upgrade with UV lighting.

Regulating Humidity: More Than Just Comfort

Humidity: Finding That Goldilocks Zone

Let's talk humidity, the often-overlooked diva of indoor comfort. Too much, and you're navigating the Amazon rainforest from your couch. Too little, and it's a Sahara Desert staycation. But did you know that extremes in humidity can have health implications too? Here's the scoop: High humidity promotes mold growth and dust mites – a true enemy for allergy sufferers. On the flip side, dry air can leave our skin cracked, make our eyes itch, and even irritate our respiratory tracts. For those of you with musical inclinations, it can play havoc with wooden musical instruments! The key is balance. Finding that sweet spot, around 40-60% humidity, is essential for both our well-being and comfort. So, DIY aficionados, consider this a call to action. It's time to dive into the world of humidity control!

Balancing Act: Your HVAC's Role in Humidity Control

Picture this: A balmy summer afternoon outside, yet you're inside, comfortably sipping your coffee, oblivious to the muggy mess just beyond your windows. How? Thank HVAC! These systems are more than temperature titans; they're humidity heroes. When air is cooled by your HVAC, moisture is extracted, acting as a dehumidifier. On the flip side, during drier months, HVAC can work with humidifiers to add needed moisture back into the air.

For those diving into the DIY realm, understanding the synergy between HVAC and humidity is paramount. Choose the right HVAC, and it does more than cool or heat — it balances. Just like finding that sweet spot in a comfy armchair, your HVAC finds and maintains the "just right" zone of humidity for optimum indoor comfort. Cheers to fewer frizzy hair days and perfect room ambiance!

Navigating the Humidity Tightrope: The Unseen Perils of Imbalance

Walking a tightrope requires balance. Similarly, maintaining the right humidity is essential, not just for comfort, but for health. Tip too much one way, and you're stepping into mold's favorite playground. Mold spores adore damp environments, swiftly turning cozy nooks into potential health hazards. And it's not just a visual nuisance; mold can lead to respiratory problems and allergic reactions.

Lean too far the other way with dry conditions, and you're rolling out the red carpet for increased pathogens. Without enough moisture, our respiratory system's natural defenses can weaken, making us more susceptible to colds and flu. Plus, dry skin and static shocks? Ouch!

So, while diving into the DIY universe, remember: controlling humidity isn't just about comfort. It's an essential step in maintaining a safe, healthy living environment.

Proper Ventilation: The Breath of a Building

Breathing Easy: The Fresh Air Elixir for Indoors

Ever enjoyed the invigorating feel of a brisk walk outdoors? That's nature's own blend of fresh air working its wonders. Now, imagine bottling that up for your living space. Homes and offices can often become stagnated cocoons, unwittingly harboring airborne nasties. Enter the savior: fresh air exchange.

Allowing for a controlled dance between outdoor and indoor air ensures that pollutants don't overstay their welcome. Without this exchange, we risk trapping VOCs, carbon dioxide, and other contaminants, turning our cozy corners into unintentional petri dishes. An efficient fresh air exchange system replenishes oxygen levels, dilutes indoor pollutants, and balances moisture levels – a triad of health boons.

For the DIY enthusiast, it's more than just décor and layout. It's about understanding the invisible currents that shape wellness indoors. Remember, a well-ventilated space isn't just refreshing; it's a health non-negotiable.

Crafting Clean Spaces: HVAC as Your Indoor Air Guardian

The DIY enthusiast knows that comfort isn't just about plush cushions and cozy nooks. Air quality plays a starring role. That's where our trusty HVAC systems come into the limelight. Think of HVAC not just as temperature managers, but as gatekeepers of indoor air purity.

HVAC systems go beyond heating and cooling; they champion ventilation. By constantly circulating air, these systems dilute indoor pollutants, ensuring a decrease in their concentration. Modern units are even designed to strategically exchange indoor air with fresh outdoor air, pushing out stale, contaminated indoor atmosphere and welcoming a breath of rejuvenation.

In essence, every time your HVAC whirls to life, imagine it's doing a sweeping dance, catching unseen pollutants and gently escorting them out, ensuring your indoor spaces stay as refreshing as a morning breeze. A clean home is more than what meets the eye—it's what fills the lungs!

The Hidden Snag in "Sealed Tight" Structures

For DIY enthusiasts who pride themselves on creating well-insulated and sealed spaces, there's a hidden snag to be wary of: the risk of overly 'tight' buildings. Now, while impeccable insulation can do wonders for energy efficiency, it might inadvertently seal in more than just warmth.

Without proper ventilation, these sealed environments can trap pollutants, creating a soup of stale air and contaminants that has nowhere to go. From volatile organic compounds released by fresh paint or new furniture, to the simple carbon dioxide we exhale, the indoor air cocktail can quickly become hazardous.

Moreover, trapped humidity can invite unwanted guests like mold and mildew. Not only do these fungi love such conditions, but they can also wreak havoc on health and building structures. The lesson? Balance is key. While sealing tight saves energy, ensuring proper ventilation preserves the purity and safety of indoor atmospheres.

Beyond Physical Health: Psychological and Emotional Well-being

Air Quality: More Than Just a Breath of Fresh Air!

For many DIY enthusiasts, the focus is often on visible improvements. Yet, the air we breathe can have invisible effects, especially on our mental well-being. Quality air isn't just about avoiding coughs or sneezes; it's intricately linked with our brain health.

Ever noticed feeling sluggish or foggy in a stuffy room? Poor air quality can decrease cognitive performance, leading to reduced concentration and mood dips. Prolonged exposure to indoor pollutants can exacerbate anxiety and depression symptoms. On the brighter side, fresh air has the opposite effect, enhancing clarity and uplifting our spirits. Imagine the joy of reading a book or doing a puzzle in a well-ventilated space!

Prioritizing air quality isn't just a physical health move. It's a nod to our mental wellness, proving once again that what's unseen can be just as vital. Breathe in, breathe out, and let your mind flourish!

Elevate Your Mindset: Regulate the Climate!

Can you believe something as simple as temperature regulation and clean air profoundly influences our mental equilibrium? It does!

Maintaining an optimal room temperature isn't just for physical comfort; it's a game-changer for focus and productivity. Ever tried concentrating in a sweltering or freezing room? It's near-impossible! Balanced temperatures set the scene for a Zen-like concentration zone, enabling our minds to function at peak levels.

Meanwhile, clean air does more than keep us sneeze-free. Breathing unpolluted air revitalizes the brain, clarifying thought processes and buoying mood. Think of it as a cerebral detox!

Dive deep into HVAC know-hows not just for a comfortable abode, but for a sanctuary that rejuvenates the mind. Elevate your living space; elevate your mindset!

The Science of Comfort: Boosting Productivity & Health!

In the vibrant world of DIY, we often emphasize the visual and tactile aspects of our spaces. Yet, what if we told you that beneath the surface, the invisible realm of air quality has a monumental role in shaping our daily lives? Dive deep with us!

Research has provided fascinating insights into the correlation between well-ventilated spaces and the productivity of their occupants. One might think it's about comfort, but it's more than that. Proper airflow and temperature regulation can actually spark our cognitive abilities, fueling efficiency and creativity. Furthermore, studies have unveiled a golden nugget: environments with balanced temperatures and clean air report fewer sick days among their inhabitants! It's a win-win situation.

Embrace the art and science of HVAC. Beyond aesthetics and comfort, it's about cultivating spaces that actively contribute to our well-being and success. Craft, breathe, thrive!

The Silent Guardian of Our Health

The Ultimate Harmony: HVAC's Symphony for Holistic Wellness

Step aside décor and furnishings; there's an unsung hero in the realm of DIY, working tirelessly to ensure our holistic well-being: HVAC! Let's delve into this revelation.

The magnificence of HVAC systems lies not just in temperature adjustments but in their comprehensive impact on our lives. Imagine a trifecta of benefits: Physical – ensuring we breathe unpolluted air, dodging allergens, and pathogens. Psychological – sculpting an environment where focus and productivity soar. Emotional – fostering spaces where comfort makes us feel secure and cherished. Every hum of the air conditioner, every whisper of a vent, it's all contributing to a symphony of wellness. HVAC isn't just hardware; it's the heart ware of our homes and offices. In the DIY world, it's essential to acknowledge and applaud this game-changer. Embrace it, and let holistic well-being resonate in every corner!

Rise to the Occasion: Championing Your HVAC's Noble Cause

Here's your rallying call: Our beloved HVAC systems are not mere tools but guardians of our living spaces. Their quiet diligence keeps us in a sweet spot of comfort, health, and efficiency. But remember, even heroes need check-ups! Regular maintenance is essential not just for longevity, but for peak performance. Dive deeper, and familiarize yourselves with its intricate functionalities; there's empowerment in understanding. Beyond just care, we must also champion the boundless innovations within the HVAC world. The sector is always evolving, striving for greener, smarter, and more efficient systems. So, as stewards of our spaces, let's ensure our HVAC systems never falter in their duty. Prioritize care, celebrate advancements, and most importantly, stay engaged. For in doing so, we safeguard not only our homes and offices but also our planet.

CHAPTER 2
UNDERSTANDING HVAC COMPONENTS AND THEIR ROLES

Thermostats

The Thermostat: Your Personal Climate Control

Ah, the thermostat! This nifty device is your go-to commander for indoor comfort. Think of it as the remote control of your home's climate. It constantly keeps an eye (or sensor) on the ambient temperature, ensuring it's not too chilly for those cozy movie nights or too warm when you're busy in the kitchen. By simply adjusting this gadget, you dictate how warm or cool your space should be. It's not just about convenience; a properly set thermostat can be your ally in energy efficiency, helping save a few bucks on bills. So next time you walk past your thermostat, give it a nod of appreciation; it's tirelessly working to make sure your indoor oasis is just as you like it!

The Thermostat: The Unseen Maestro of Comfort

Let's imagine your HVAC as a grand orchestra, playing a symphony of warm and cool breezes. Now, who do you think is the conductor? That's right, the thermostat! This unassuming device doesn't just display numbers; it's the maestro, ensuring each section (or component) of your HVAC plays in perfect harmony. By maintaining the temperature you've set, it commands the cooling to hush when it's too chilly or beckons the heat during a cold draft. This not only guarantees comfort but also prevents the system from overworking, thereby ensuring efficiency and prolonging its lifespan. So next time you adjust that dial or tap that screen, remember: you're communicating with the genius brain behind the climate magic in your abode!

Furnaces & Heaters

Warm Embrace: Meet Your Heating System

Curling up in a cozy blanket on a chilly day is pure bliss, isn't it? Well, the unsung hero behind that snug indoor environment is your heating equipment. It's like a dedicated kitchen chef, but instead of whipping up dishes, it cooks up warmth! Your heating system takes in cold air, adds a dash of heat through its special 'oven' (often a furnace or boiler), and sends it out to every corner of your home, ensuring that the chills don't stand a chance. Just as a chef uses various methods to cook, there are different types of heaters – from furnaces to heat pumps. Regardless of the type, the end goal remains the same: to keep you warm, comfy, and away from those winter woes.

Inside the Heating Magic: Key Components Decoded

Pop open the hood of your car, and you'll find a world of components working in harmony. Similarly, your heating system is an orchestra of parts, each playing its role to perfection. Let's break it down:

The Burner: Think of this as the musician setting the beat. It initiates the heating process by combusting the fuel.

Heat Exchanger: The maestro, taking heat from the burner's combustion and transferring it to the air. A vital component ensuring the warmth is just right!

Blower: Your very own wind machine. Once the air gets heated, this guy steps in, sending that warm air through your vents, ensuring every room feels like a cozy embrace.

Flue: The responsible exit artist. It safely guides the combustion gases out of your home, keeping the environment clean.

The Trio of Warmth: Understanding Different Heaters

When the winter winds howl, we all have a favorite sweater, but what's the ideal heater for your home? Let's break down the band:

Gas Furnaces: The popular kid in town. Efficient and often less costly in terms of operating expenses. They use natural gas to create a cozy environment and are known for their quick heating prowess.

Electric Heaters: The eco-conscious artist. No fumes, no combustion – just a quiet and clean operation. They might have a higher operating cost but shine in areas without natural gas lines.

Oil Furnaces: The classic rocker. Older homes might still sport these, using oil to produce heat. They require a tad more maintenance but can churn out a lot of heat!

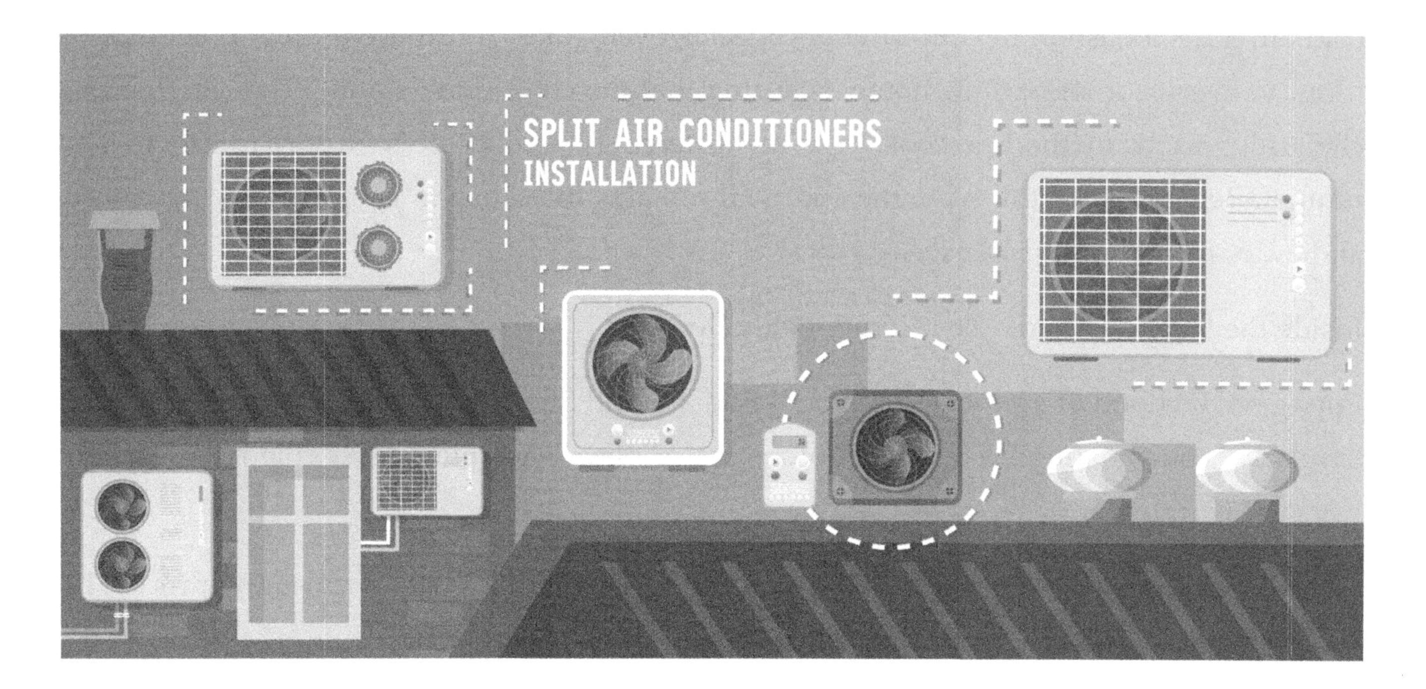

Air Conditioners & Cooling Units

Keeping Your Cool: Air Conditioners Decoded

Ever wondered about the unsung hero during those sweltering summer months? Enter: The Air Conditioner. This marvel doesn't just blow cool breezes your way; it operates on some nifty science! Imagine your home as a hot, crowded dance floor. The air conditioner is the bouncer that escorts the unwanted heat outside, ensuring the inside remains refreshingly chill. In layman's terms, it takes in warm room air, works its magic to strip away the heat, and then sends the cooled air on a delightful journey back into your living space. The result? Your sanctuary stays crisp, letting you bask in comfort, even when the world outside feels like a sauna. So, the next time you're sipping a cool drink indoors, take a moment to toast to your AC – the real MVP of summer!

The Cooling Quartet: Decoding the Core Components of Air Conditioners

Ready for a deep dive into the chill zone? Let's unbox the magic ensemble that orchestrates those summer breezes indoors!

Compressor: The captain of the ship! It pressurizes the refrigerant, setting it on its heat-exchanging journey. Think of it as getting the refrigerant prepped and pumped for its mission.

Coils: The stages where all the action happens. As refrigerant courses through, these coils play a game of hot-and-cold, absorbing indoor warmth and releasing it outside.

Refrigerant: The star performer! This fluid wizard transforms from gas to liquid and back, absorbing and shedding heat in the process. It's the heart and soul of the cooling cycle.

Fans: The diligent promoters, spreading the cool vibes. Once the air is chilled, fans ensure it circulates, giving your space that fresh, comfortable embrace.

Picking Your Cool Crew: Different Types of Air Conditioners Demystified

Venturing into the AC world feels like walking into an ice cream parlor. Choices galore! Let's make your choice crystal clear:

Central Air Conditioners: The big-league player! This system is all about unified cooling, using ducts to spread chill vibes throughout your entire home. Perfect for those who adore consistent coolness in every corner.

Ductless Mini-Splits: Sleek, stylish, and super-efficient. These guys offer targeted cooling for specific zones or rooms. No ducts? No problem! Mini-splits are perfect for additions, converted spaces, or homes without pre-existing ductwork.

Window Units: The classic maestros. Compact and perfect for single rooms. Sit them on your window ledge, plug them in, and enjoy the breezy serenade.

Ductwork & Vents

Ductwork Demystified: The HVAC Highway

Navigating the world of HVAC can sometimes feel intricate, but let's simplify a core element: the ductwork. At its essence, ductwork is the architectural marvel within your walls, floors, and ceilings that acts as the transportation hub for your HVAC system. Much like the circulatory system in our bodies, these ducts distribute cold or hot air, ensuring consistent temperatures throughout a building.

Every time you set your thermostat, it's the ductwork that answers the call, efficiently channeling the conditioned air from the HVAC unit to the intended rooms. Precision is key here. The design and layout of these ducts must be spot on to guarantee optimal airflow, preventing hot or cold spots in different areas.

While they might remain hidden and often overlooked, ducts are truly the backbone of your HVAC system. Proper installation and regular maintenance of this "air highway" are paramount. After all, a well-maintained duct system is the silent hero behind your everyday comfort, making sure that every room in your space is just right, regardless of the season.

The Vitality of Duct Maintenance: Airflow and Quality

Ducts, while often out of sight, should never be out of mind. Their maintenance is a cornerstone for an efficient HVAC system. Why, you ask? It's simple: clean ducts equal efficient airflow and pristine air quality. Over time, dust, allergens, and even mold can accumulate inside, obstructing the flow and contaminating the air you breathe. Regular cleaning prevents these unwanted guests from setting up camp.

But it's not just about cleanliness. Maintaining the integrity of your ducts by checking for leaks or damages ensures the HVAC system doesn't work overtime. This translates to energy savings and prolonged system life. Moreover, with unobstructed and intact ducts, you're guaranteed a consistent temperature throughout your space, enhancing comfort.

In short, think of duct maintenance as a wellness checkup, vital for the health of your HVAC and, by extension, for you and your building's occupants.

Air Handlers & Blower Units

The Heartbeat of HVAC: Deciphering Air Handlers

An HVAC system is an intricate puzzle, each piece pivotal to the overall picture of comfort. At its core, guiding this process, is the air handler. What's its role? Let's dive in.

An air handler is, in essence, a cabinet containing essential components that move and regulate the air within the HVAC system. Think of it as the control hub, directing traffic and ensuring every component does its job in harmony.

Within its structure, the air handler hosts a blower, which is responsible for circulating air throughout the building. But it doesn't stop there. Attached to this blower are filters that trap pollutants, ensuring the air that circulates is as pure as can be. In some cases, air handlers might also contain heating or cooling elements, making them even more integral to temperature regulation.

Positioning is crucial. Typically, they're found in basements, attics, or specially dedicated closets. Their placement often depends on the system's design and the building's structure. No matter where they are, their objective remains unchanged: to regulate and circulate air efficiently.

Maintenance is, as with all HVAC components, vital. Regular checks and cleaning ensure they function optimally. In turn, an efficient air handler translates to a well-performing HVAC system, which guarantees a comfortable indoor environment.

To sum it up, while the air handler might be just one component in the vast HVAC orchestra, its role is undeniably central. Understanding its function and importance is key to appreciating the symphony of comfort it orchestrates daily.

Circulation Central: The Role of Air Handlers in Temperature Maintenance

An air handler's role might seem straightforward, but it's a task of paramount importance. Often regarded as the heart of the HVAC system, it ensures every room in a building maintains the desired ambiance.

At its essence, an air handler consistently circulates air, which is fundamental in achieving and maintaining the set temperature. The mechanism behind this? It's a blend of technology and precision. When the thermostat signals a need for cooling or heating, the air handler leaps into action. It takes in air, passes it through filters to remove any contaminants, and then pushes this conditioned air throughout the building.

Moreover, the consistent circulation guarantees no room is left untouched. Whether it's the sunlit living room in need of cooling or the chilly basement yearning for warmth, the air handler ensures uniformity.

Its function is integral not just for comfort but also for efficiency. Without its tireless work, other components might overstrain, leading to energy wastage. So, when you enjoy that perfect room temperature, remember it's the air handler working diligently behind the scenes.

Filters

Clean Air Commanders: Understanding Air Filters

Air filters are an indispensable element within the HVAC system. Their primary function? To ensure the air circulating within your home or building remains clean and free from pollutants.

Let's break this down. As the HVAC system works to regulate temperature, it draws in air from the environment. This incoming air often carries with it a variety of particulate matter such as dust, pollen, and sometimes even bacteria. Enter air filters. Positioned strategically within the HVAC system, they act as the first line of defense against these contaminants.

Constructed using fibrous or porous materials, these filters are designed to trap and hold onto the pollutants, allowing only clean air to pass through. The result? An environment that not only feels comfortable temperature-wise but also promotes health and well-being by reducing the risk of allergies and respiratory issues.

Regular maintenance of these filters is crucial. Over time, as they trap more and more pollutants, they can become clogged. Ensuring they are cleaned or replaced periodically guarantees that the HVAC system operates efficiently and continues to provide a steady supply of clean, breathable air. It's the unseen guardian ensuring every breath you take is a healthy one.

The Vital Role of Air Filters: Breathing Clean

The quality of indoor air is paramount to our health and comfort, and air filters play a pivotal role in ensuring this. While HVAC systems are responsible for maintaining our desired indoor temperature, it's the air filters that take charge of the purity of the air we breathe.

Every day, a multitude of particles, including dust, pollen, and microorganisms, aim to make their way into our living or working spaces. Air filters act as barriers, capturing these contaminants and preventing them from circulating within the indoor environment. This function is particularly vital for those who suffer from allergies, asthma, or other respiratory conditions, ensuring that the air remains free of potential triggers.

Beyond health benefits, clean air contributes to a more pleasant living or working environment. Stale or polluted air can lead to discomfort, fatigue, and reduced concentration levels. With efficient air filters, there's an assurance of a fresher, cleaner atmosphere.

Regularly checking and replacing air filters is not just about equipment efficiency; it's about prioritizing health and ensuring an optimal indoor environment. In essence, they are silent workers that uphold the promise of a cleaner, healthier indoor life.

Evaporator & Condenser Coils

Coils: The Essential Change Agents in HVAC

Diving into the core mechanics of the HVAC system, we come across the indispensable coils. These components, often made of copper, aluminum, or steel, serve a vital role in the operation of your heating and cooling system.

Coils come in two primary types: evaporator coils and condenser coils. Evaporator coils work by absorbing heat from the indoor air, allowing the refrigerant inside them to evaporate. This transformation from a liquid to a gas extracts heat, thereby cooling the surrounding air.

On the flip side, condenser coils release the absorbed heat to the outside environment. Here, the refrigerant changes back from its gaseous state to a liquid, releasing heat in the process.

In essence, coils are the central hubs where the magic of temperature transformation takes place. Their efficient operation is crucial for the HVAC system to provide the desired indoor climate. By understanding their function, one gains a clearer insight into the intricate dance of heating and cooling in any space.

Coils in Action: The Cool Cycle Explained

The cooling process in your HVAC might seem complex, but when broken down, it's all about the clever choreography between the evaporator and condenser coils. The evaporator coil plays a pivotal role in absorbing heat. As the refrigerant flows through this coil, it changes from a liquid to a gas. This phase change pulls heat from the surrounding air, delivering a fresh, cool breeze to your space.

Conversely, the condenser coil is tasked with the opposite: expelling this absorbed heat to the outdoors. Here, the refrigerant, now in its gaseous state, flows through the coil and turns back into a liquid, releasing the trapped heat in the process.

These coils, working in tandem, maintain the balance of your indoor climate. Their cyclical actions—absorbing and releasing heat—are the very heartbeat of the cooling process in any HVAC system.

Refrigerants

Understanding Refrigerants: The HVAC's Lifeline

Refrigerants are fundamental to the functionality of an HVAC system. These are specialized chemical compounds designed to transition between gaseous and liquid states at specific temperatures and pressures. This ability to change phases efficiently is crucial for the heat absorption and release processes within the HVAC system.

As the refrigerant circulates, it absorbs heat from the indoor air, causing it to evaporate and become a gas. As it moves through the system, particularly through the compressor and the condenser coil, it releases this heat to the outdoors and returns to a liquid state. This cyclical movement of refrigerants, transitioning between liquid and gas, is the driving force behind the heating and cooling we experience.

In essence, refrigerants are the circulatory system of your HVAC, ensuring the transfer of heat and enabling our homes and offices to remain comfortable, regardless of the external weather conditions.

The Role of Refrigerants: Masters of Heat Transfer

The core role of refrigerants in an HVAC system is pivotal — they are the champions of heat absorption and release. When we talk about the cooling process, refrigerants are at the very heart of it. As they circulate through the system, they undergo phase changes, shifting from liquid to gas and back again.

During the cooling cycle, the refrigerant, in its gaseous form, absorbs heat from the indoor environment, causing it to evaporate. This heat-laden gas is then compressed and conveyed to the condenser coil, where it releases the absorbed heat to the outside environment, returning to its liquid state. This continuous cycle of absorbing and releasing heat is what gives us the cooling comfort we desire.

In short, without refrigerants doing their crucial job, our HVAC systems would be unable to provide the refreshing, cool environment we've come to rely on during warm seasons.

CHAPTER 3

KEY COMPONENTS AND THEIR FUNCTIONS

Understanding the inner workings of the HVAC system is crucial for any homeowner or technician. Each component has its unique role, and when they all function in harmony, we get a seamlessly operating system. Here's a detailed look:

Thermostats

Thermostats: Your Climate Control Panel

Thermostats are the unsung regulators of our comfort zones. Acting as the main interface between you and your HVAC system, they gauge the surrounding temperature and respond to your climate preferences. By displaying the ambient temperature, thermostats let you adjust settings to your desired level, whether that's a cozy warmth on a chilly evening or a refreshing coolness during a midsummer day. These devices work in tandem with the HVAC system, sending signals to either crank up the heat or dial it down, all based on your selections. Their precision ensures that your living or working space maintains the optimal temperature, delivering comfort and efficiency. So, when you think of thermostats, consider them as the trusted captains of your indoor climate ship.

The Essential Mechanism: Thermostat Functions

Dive into the world of HVAC, and you'll soon recognize the pivotal role of thermostats. Acting as the central control panel, the thermostat is the interface through which all comfort decisions are made. It vigilantly monitors the room temperature, comparing it with the user-defined settings. When there's a deviation, it springs into action. If the room's too chilly, it signals the heating system to kick into gear. Conversely, if things get a tad warm, it directs the cooling system to spring forth. Beyond just monitoring, it efficiently governs the start and stop of the HVAC components, ensuring that energy is utilized judiciously. This balance between user comfort and system efficiency is what makes thermostats invaluable. In essence, they are the diligent guardians of your indoor climate, always striving for perfection.

Heating Units

Decoding Heaters: The Warmth Givers

Delve into HVAC, and heaters emerge as indispensable champions. These equipment pieces are laser-focused on one primary mission: raising the temperature of indoor air. As the mercury dips and cold starts nipping at our toes, heaters work diligently, turning cold drafts into comforting warmth. While there are various types of heaters available, their core function remains consistent - to elevate indoor temperatures and bestow comfort. By using energy sources such as gas, electricity, or oil, they effectively convert that energy into heat. In the grand scheme of HVAC, understanding heaters is paramount. They stand as the bulwark against winter's chill, ensuring homes and offices remain welcoming havens no matter the external conditions.

Unpacking the Heater: Key Components Demystified

Every heater, despite its differences, operates with a set of fundamental components at its core. The *Burner* jumps into action first, igniting the fuel and initiating the heating process. Then, the *Heat Exchanger* steps in, expertly transferring the heat from combustion to the air destined for your rooms. The *Blower* doesn't stay behind; it ensures the newly-warmed air circulates throughout your space efficiently. Lastly, the *Flue* plays a pivotal role in safety. It directs combustion gases out and away from the indoor environment. Together, these components form a well-coordinated team. Understanding their individual roles offers insights into the seamless operation of heaters, providing consistent warmth every time.

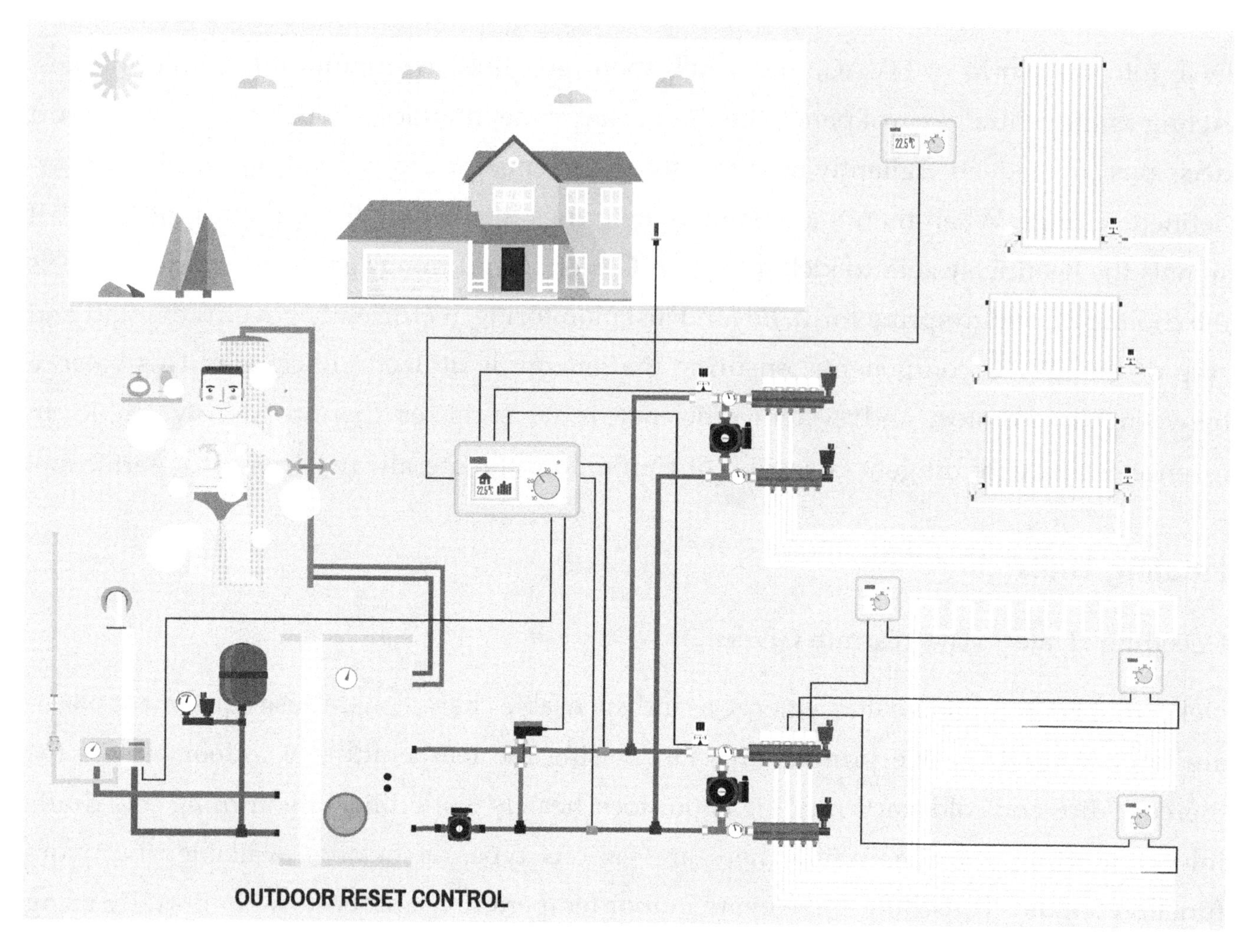

Heater in Action: Understanding its Core Functions

At the heart of every heater lies a simple yet crucial mission: to envelop your space in comforting warmth. First, it deftly converts fuel into heat, a transformative process where energy sources become tangible warmth. But it doesn't stop there. The heater then elevates the air's temperature, preparing it for its journey through your living or working space. And to ensure every corner gets its share of warmth, the heater circulates this heated air methodically and uniformly. In essence, from initiating the heating process to the final spread of warmth, the heater's function is systematic, ensuring you're cocooned in the perfect temperature every time.

Cooling Units

Welcome to the refreshing realm of cooling devices. Simply put, these are machines designed with one primary purpose: to reduce your indoor air temperature. Think of them as your indoor climate champions, standing guard against the sweltering summer sun. These devices achieve their goal through a series of processes, including removing excess heat and moisture from your indoor environment. At the core, their function is all about balance - ensuring your spaces remain comfortable, no matter how blistering the outdoor conditions might be. Knowledge of these systems is your first step towards enjoying a cooler, more pleasant indoor space.

Components: The Building Blocks of Cooling

Dive into the anatomy of your cooling device and you'll find some impressive components working in concert. First up is the compressor, the heart of the system, which pumps the refrigerant, our cooling agent. Then we have the coils, which play a pivotal role in heat exchange. The refrigerant, ever crucial, travels through these coils, absorbing and releasing heat as it shifts between liquid and gaseous states. Last but not least, fans aid in distributing the conditioned air evenly throughout your space. Together, these elements form the orchestra of your cooling symphony, ensuring optimal performance every time.

Function: The Science Behind the Cool

Central to your cooling system's operation is the principle of heat transfer. Its main task? Extracting warmth from the indoor air and dispatching it outdoors. Here's how it's done: the system pulls in hot indoor air, which then passes over cold coils. This process siphons off the heat, leaving the air cooler. Afterward, this freshly chilled air is propelled back into the room while the captured heat is jettisoned outside. It's a continuous cycle of heat extraction and expulsion, ensuring your living spaces remain refreshingly cool and comfortable.

Air Handlers

Definition: The Heartbeat of Air Movement

At the very core of an HVAC system, there lies a pivotal unit: the air handler. It's not about merely pushing air around. Instead, it ensures that this air moves with purpose, regulation, and direction. An air handler delicately balances the indoor environment by regulating airflow, ensuring that every space in a building receives its fair share of treated air. Whether it's a comforting warmth on a chilly evening or a refreshing breeze on a hot afternoon, these units manage the ebb and flow of your indoor climate with unmatched precision.

Function: The Unwavering Sentinel of Comfort

The pivotal role of the air handler goes beyond just moving air; it's about maintaining equilibrium. By ensuring a consistent and well-regulated airflow, the air handler stands as the guardian of your indoor comfort. No matter the season, it responds adeptly, adjusting the volume and direction of air to ensure your set temperature remains stable. In essence, the air handler acts as the reliable backbone, bridging the gap between your desired comfort level and the HVAC system's capability to deliver just that. A well-functioning air handler is synonymous with an unwavering indoor climate.

Ductwork

Definition: The Essential Arteries of HVAC

Ductwork: It's not just a series of pipes; it's the lifeline of your HVAC system. Acting as the principal conduit, ductwork meticulously channels cold or hot air to each part of your building, ensuring comfort in every corner. Just as our veins and arteries transport blood throughout our bodies, the ducts carry conditioned air, vital for a building's comfort. With such a critical role, understanding the importance and intricacies of this "air transportation system" is crucial for anyone keen on HVAC mastery.

Function: The Symphony of Air Distribution

Picture your HVAC as an orchestra, and the ductwork as its conductor. Its primary role? To guide the harmonious flow of conditioned air to every nook and cranny of your building. This systematic distribution ensures that every space, be it a cozy corner room or a sprawling hall, gets its fair share of warmth or coolness. Properly designed and installed ducts eliminate hot or cold spots, ensuring a consistent, comfortable indoor environment. So, next time you feel that gentle gust of perfect air, remember the conductor – your ductwork – making it all happen.

Air Filters

Definition: The Unsung Sentries of Indoor Air

When discussing HVAC, air filters and purifiers might not always steal the limelight, but they play a pivotal role in keeping our air clean. Think of them as the vigilant gatekeepers of your indoor environment. Their mission? To capture and contain those unwanted airborne guests - from pesky dust particles to harmful pollutants. These components act as a sieve, ensuring the air you breathe remains uncontaminated and fresh. In the vast world of HVAC, these elements stand as silent champions, consistently guarding against invisible threats.

Function: Breathing Easy with Every Cycle

In the grand ballet of HVAC operations, air filters and purifiers dance with grace, ensuring each gulp of air we inhale is as pristine as nature intended. Their role? Simple but vital. As the HVAC system operates, they diligently sift through the circulating air, trapping unwanted particles like dust, pollen, and even microscopic allergens. The result? An indoor atmosphere that's not just comfortable in temperature but also in purity. So, every time your system hums to life, remember it's not just about warmth or coolness; it's about clarity of breath.

Coils (Evaporator and Condenser)

Definition: The Ebb and Flow of Temperature

When diving into the core mechanisms of an HVAC system, there are specialized zones where real magic happens: the areas where refrigerants shift their states. These parts, crucial pivot points in temperature control, allow the refrigerant to either evaporate or condense. Think of them as the "chefs" in the HVAC kitchen, skillfully controlling the "recipe" of your indoor climate. Through their operations, these zones facilitate the efficient transfer of heat, orchestrating the perfect indoor ambiance for your comfort.

Function: The Core of Cooling

The evaporator and condenser are integral components of your HVAC system. Their principal function revolves around managing the phase change of the refrigerant. The evaporator absorbs heat from the indoor air, causing the refrigerant to evaporate.

Conversely, the condenser releases that absorbed heat to the outdoors, returning the refrigerant to a liquid state. This systematic phase change, facilitated by these components, is pivotal in efficiently transferring heat from inside your home to the outside environment, ensuring a consistently cooler indoor atmosphere.

Refrigerants

Definition: The Heartbeat of the System

Refrigerants are specialized chemical compounds coursing through your HVAC system. Think of them as the bloodline of your cooling mechanism. Their ability to change from liquid to gas and back again under controlled conditions allows the HVAC system to absorb, transport, and release heat. Their unique properties, tailored for these phase transitions, make them indispensable in ensuring your space remains at your desired temperature. When it comes to HVAC, these compounds aren't just chemicals; they're the unsung heroes maintaining your comfort.

Function: The Secret Behind Cool Comfort

At the heart of your HVAC system's prowess is the refrigerant, the silent workhorse. It gracefully dances between liquid and gas phases, absorbing and releasing heat like a seasoned performer. This intricate ballet of heat transfer is what allows our homes and offices to feel like sanctuaries, regardless of the blazing sun or sticky humidity outside. Simply put, without refrigerants doing their heat-exchange magic, our modern quest for perfect indoor climates would be just a dream.

CHAPTER 4

CHOOSING THE RIGHT HVAC SYSTEM FOR YOU

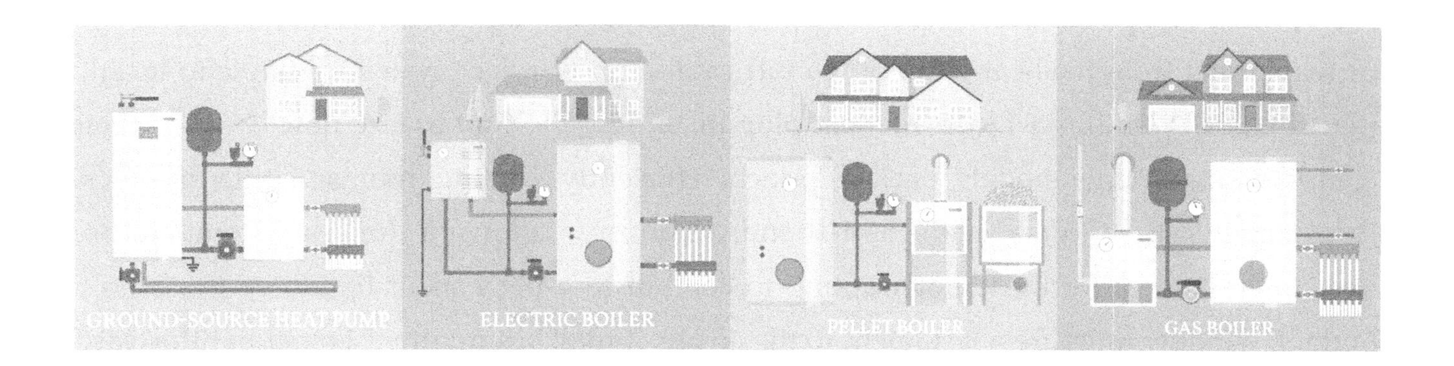

Understanding Different Types:

Central HVAC Systems: Comprehensive Climate Control

Central HVAC systems stand as a testament to efficiency and unified temperature management for large spaces. Unlike individual room-based solutions, a Central HVAC system operates from a singular location, typically centralized within a building. It circulates conditioned air—be it warmed or cooled—through a series of ducts that distribute it to various rooms or zones. This ensures a consistent temperature throughout, without the discrepancies sometimes noted in buildings using multiple standalone units. Given its capacity to handle larger areas and its streamlined approach to climate control, the Central HVAC system has become an integral choice for many residential and commercial structures.

Ductless Mini-Splits: Flexibility Meets Function

Ductless mini-splits are the unsung heroes for those spaces where ductwork might seem impractical or unwarranted. Operating without the maze of ducts traditional HVAC systems rely on, these systems provide individualized temperature control, room by room. Comprising an outdoor unit and one or several indoor units, installation is straightforward: just mount, connect, and power on! Especially ideal for additions, renovated spaces, or older buildings, they eliminate the need to extend or add ductwork. Plus, the zoned approach of mini-splits lets you cool or heat specific areas, ensuring energy isn't wasted on unoccupied spaces. With easy installation, energy efficiency, and pinpoint climate control, ductless mini-splits are the savvy solution for localized comfort.

Window Units: Compact Cooling Powerhouses

Window units, often a godsend for apartments and smaller spaces, offer a blend of convenience and portability. These self-contained systems snugly fit into window frames and provide relief from the heat without the intricate setup of larger systems. Simple to install, they're the DIY enthusiast's dream: just plug in, and let the cool breeze flow! Despite their compact size, many models pack a punch, efficiently cooling rooms within minutes. Additionally, they come with user-friendly controls, allowing for easy temperature adjustments. Ideal for those not wanting to commit to a permanent fixture or those on a budget, window units are a testament to the adage, "Small but mighty." For straightforward cooling without the fuss, window units are your go-to.

Portable Units: Cooling on the Go

Portable units are the epitome of flexibility in the HVAC world. As the name suggests, these units aren't tied down; they move as you do. Featuring wheels at the base, they can be effortlessly transported from one room to another, targeting the cooling wherever it's needed the most. A favorite among renters and those with ever-changing spatial needs, setting them up is a breeze — just plug into an outlet and vent out via a window. No need for complex installations or commitments! Plus, their efficiency doesn't take a backseat. They deliver ample cooling, keeping spaces comfortable during those sweltering days. If you're looking for an adaptable solution that caters to dynamic lifestyles, portable units are your ticket to instant relief.

Hybrid Systems: The Best of Both Worlds

Step into the future with hybrid systems, an innovative answer to modern HVAC needs. Marrying the efficiency of heat pumps with the raw power of traditional furnaces, these systems are a game-changer. When it's mild outside, the electric heat pump takes the lead, efficiently warming or cooling your interiors. But when Jack Frost really starts nipping, the furnace fires up, ensuring you remain cozy no matter the chill. It's a harmonious switch, all aimed at energy conservation and keeping those utility bills in check. Think of it as a tag-team duo where each player shines in its moment. For the eco-conscious homeowner keen on flexibility without compromising on comfort, the hybrid system is the green flag you've been waiting for.

Assessing Your Specific Needs:

Space Considerations:

Size and Layout of Your Home: The Blueprint to Your Comfort

Your home's size and layout are like the DNA of your HVAC needs. Not all homes are built equal, and neither should their heating or cooling solutions be. Begin by gauging the square footage; a more spacious abode might call for a robust unit. But size isn't everything. The design, with its nooks, crannies, and open spaces, matters too. A sprawling ranch-style home has different demands than a multi-story townhouse. And remember, sometimes it's not about sheer power but smarter distribution. Zone systems can be the knight in shining armor for homes with varied spaces, ensuring each area gets the TLC it deserves. So, before diving into HVAC choices, get intimate with your home's dimensions and design. It's the starting point of a journey to impeccable indoor comfort.

Existing Infrastructure: The Silent Determiner

Peek behind the walls, and you might find your home telling its own HVAC story. Existing ductwork can be a goldmine or a challenge, depending on its state. If your home boasts a duct network, it's like having half the puzzle completed. Systems that rely on ducts—like certain central HVACs—slide in seamlessly. On the flip side, if the ducts are outdated or poorly maintained, think twice. They might require renovation or even replacement. Homes without any ductwork offer a fresh canvas, making ductless systems, like mini-splits, an attractive option. The lesson? Know your home's backstage. Recognizing the infrastructure you already have in place can make your next HVAC decision not just good, but absolutely perfect.

Climate and Geography:

Local Climate: More Than Just Small Talk

Your local weather is more than just a conversation starter—it's a critical variable in your HVAC equation. If you're nestled in a frosty region, you'll want a heating system that doesn't just warm up a room but stands up to the coldest of nights. Meanwhile, those in hotter climes should prioritize robust cooling capabilities. The occasional chilly night? Maybe a lighter heating function will suffice. Some systems are better equipped for consistent, extreme temperatures, while others are more suited for moderate climates with fewer extremes. So, before you dive into brand names and features, pause and consider Mother Nature. After all, she has a vote in your home's comfort too.

Sun Exposure: Let the Sun Shine, But Be Ready!

While a sun-soaked room can be a mood-lifter, it's also a heat booster. Rooms graced with abundant daylight might become indoor sunbaths, requiring more vigorous cooling efforts. Conversely, those shady, tree-covered spots might remain cooler, needing less intervention from your HVAC system. It's essential to recognize these differences, as a one-size-fits-all approach won't cut it. Your south-facing sunroom and your shaded basement den have contrasting needs. As you choose an HVAC solution, factor in these sunlit variables. It'll not only improve your comfort but also optimize energy use, ensuring you aren't unnecessarily cooling or heating spaces. Remember, when it comes to HVAC, the sun's rays play a starring role.

Energy Efficiency Goals:

SEER, AFUE, and HSPF Ratings: The ABCs of HVAC Efficiency!

Dive into the world of HVAC, and you'll soon bump into these acronyms: SEER, AFUE, and HSPF. Think of them as the report cards for your HVAC system, grading its energy efficiency. Let's decode:

- SEER (Seasonal Energy Efficiency Ratio): For air conditioners, a higher SEER means more cooling bang for your energy buck.
- AFUE (Annual Fuel Utilization Efficiency): This one's for furnaces. A furnace with 90% AFUE turns 90% of its fuel into heat, wasting only 10%.

- HSPF (Heating Seasonal Performance Factor): Relevant for heat pumps, a higher HSPF indicates efficient heating.

Why care? Efficient systems reduce your utility bills and minimize your carbon footprint. So, before making an investment, check these ratings. Think smarter, not harder, and let these scores guide your HVAC choices!

Programmable Thermostats: Your Smart Energy Saver!

Step into the future of home comfort with programmable thermostats! These nifty devices aren't just modern décor pieces; they're the vanguard of energy efficiency. By allowing homeowners to set temperature schedules based on their daily routines, these thermostats ensure you're not heating or cooling an empty house.

Here's where it gets even cooler (or warmer!): integration capability with smart thermostats. Think remote access, voice commands, and even learning your preferences over time. Adjust your settings from your smartphone, or let your thermostat auto-adjust based on outside weather conditions.

The result? Substantial energy savings. By operating your HVAC system only when necessary, you're reducing wear and tear and decreasing energy consumption. It's not just about comfort, it's about being smart with your energy and your wallet. Upgrade to a programmable thermostat, and let the savings begin!

Budget Constraints:

Initial Investment: Setting the Foundation of Your Comfort

Welcome to the first big step in your HVAC journey: gauging the initial investment. It's not just about plucking a unit off the shelf; it's about understanding the nuances of cost associated with your choice.

Start with the unit's price tag. While it's tempting to gravitate towards cheaper options, remember: you often get what you pay for. Higher-end units might cost more upfront, but they tend to be more energy-efficient and durable, offering long-term savings.

Now, let's add the installation to the equation. You're not just paying for a technician's time; you're investing in their expertise. A correctly installed HVAC system runs efficiently, maintains the air quality, and, importantly, keeps those warranty terms valid.

It's vital to see this initial outlay not as an expense, but as an investment. An investment in your comfort, your home's value, and, over time, potential savings on your energy bills. So, as you ponder the digits, remember: this is the foundation of your home's comfort for years to come. Make it count!

Long-term Costs: Seeing Beyond the Present

Let's talk dollars and sense. While that shiny new HVAC system might have an attractive price tag now, it's the long run we're more concerned with. Think of your HVAC not just as a machine, but as an ongoing relationship with your wallet.

Firstly, energy consumption is the ever-present monthly guest called the utility bill. Energy-efficient systems might cost more initially, but they sip rather than guzzle power. Over months and years, these savings really add up, often offsetting the initial price difference.

Maintenance is the self-care routine of your HVAC. Regular check-ups prevent small issues from becoming big problems. While there's a cost to this, it's minor compared to potential repair bills from neglected systems. It's like gym membership for your HVAC; a little investment in its health goes a long way.

And, about those repairs—well-maintained, quality units are less prone to breakdowns. Should they occur, they're typically less severe. So, while it might be tempting to cut corners at the outset, think about the journey ahead. Sometimes, spending a bit more now can save a bundle later.

Maintenance & Upkeep:

Filter Replacements: Breathing Easier, One Change at a Time

The unsung hero of your HVAC system is undeniably the filter. This trusty component traps pollutants, ensuring cleaner, fresher air circulates in your home. But like any hero, it needs a little upkeep. Typically, filters should be replaced every 1-3 months, depending on usage and air quality. As for costs, it varies. Basic filters come cheap, but if you're eyeing those top-notch, allergen-busting varieties, expect a higher price tag. It's a small investment for breathing comfort. So, make a calendar note or set a reminder—it's a simple step that goes a long way in ensuring optimal HVAC performance

Regular Servicing: Keep Your System Humming

Think of your HVAC like a prized car; regular servicing ensures it purrs smoothly year-round. While most systems demand an annual check-up, some might need more TLC, especially if they're older or heavily used. These check-ups, conducted by professionals, can catch budding issues, extend your unit's lifespan, and maintain peak efficiency. It's more than just a once-over—it's about ensuring consistent comfort. The key is not to wait for a hiccup. Schedule these sessions like clockwork and rest easy knowing you're getting the best out of your HVAC.

Ease of DIY Maintenance: Taking the Reins

Your HVAC isn't just a machine; it's an extension of your home. While certain complexities demand professional hands, many systems empower homeowners with basic upkeep tasks. From cleaning filters to checking thermostats, you might be surprised at what you can handle! But remember, DIY doesn't mean diving in unprepared. Equip yourself with the right tools and knowledge. Check your manual, watch credible tutorials, and always prioritize safety. Still, for deeper issues or regular check-ups, a professional's touch ensures your system's longevity and peak performance?

Warranty and After-sales Service:

Duration of Warranty: Peace of Mind for the Long Haul

A warranty isn't just a slip of paper; it's your HVAC system's safety net. Understanding the duration of your warranty sets clear expectations about the timeline of guaranteed performance and protection. Longer warranties often signal manufacturers' confidence in their products. When assessing, don't just glance at the numbers. Dive into the specifics: What parts are covered? Are labor and service included? A comprehensive warranty can save significant costs and headaches down the road. Remember, it's not just about the system; it's about ensuring smooth sailing for years to come.

Coverage Details: Beyond the Fine Print

When navigating HVAC warranties, it's not just about duration. The real question is: What's covered? While some warranties might cover the system as a whole, others may be specific to certain parts like compressors or fans. Some might offer full replacements, while others could merely offer discounted repairs. Always read beyond the headlines. Does the warranty include labor costs? Are there any service fees? Knowing precisely what's included can save unexpected expenses and eliminate surprises. In the world of HVAC, being informed is your best defense against unforeseen costs.

Reputation of Manufacturer: More Than Just a Name

While a well-known name might catch your eye, delving into the reputation of an HVAC manufacturer can be enlightening. A brand's true colors shine through user reviews and expert insights. Happy customers often share their positive experiences, while disgruntled ones don't shy away from detailing issues. Look for patterns: Are there consistent complaints or praises? Moreover, consider the after-sales service—does the company respond to concerns? Stand behind its products? The reliability of a brand isn't just in its machinery but also in its commitment to customers. In HVAC choices, reputation truly matters.

Customization and Additional Features:

Zoning Capabilities: Personalized Comfort Throughout

Ever wished for a cool bedroom but a cozy living area? Enter: zoning capabilities. An HVAC feature not to be overlooked, zoning allows for distinct temperature settings in different sections of your home. Instead of a one-size-fits-all approach, you get customized comfort in each zone. With this, there's no more battling over the thermostat—a bonus for households with varied preferences. It's not only about comfort, though. Zoning can be a money-saver, as it lets you heat or cool only occupied spaces, cutting unnecessary energy use. Dive into the world of zoning, and tailor your home's environment to everyone's liking.

Integration with Home Automation: Streamlining Comfort & Convenience

With the rise of smart homes, integrating your HVAC into a home automation system is a game-changer. Picture this: arriving home to a perfectly cooled environment without lifting a finger. Or remotely adjusting the temperature when you're away. Integration ensures your HVAC system communicates seamlessly with other smart devices, optimizing comfort and energy use. It's not just about luxury; it's about elevating efficiency and convenience in modern living. By marrying your HVAC to home automation, you're not just embracing technology; you're shaping a future where your home responds intuitively to your needs.

Air Purification Features: Breathing Made Better

Clean air is not just a luxury, it's a necessity. Modern HVAC systems often come with integrated air purification features, ensuring every breath you take is free from common pollutants. These systems go beyond mere temperature control, removing dust, pollen, and other airborne irritants. With increasing concerns about indoor air quality, an HVAC with air purification isn't just an added feature; it's an ally for your health. When assessing HVAC options, consider models that boast high-grade filters or even UV treatments. In this pursuit, you're not just ensuring comfort, but promoting a healthier living space for all occupants.

Humidity Control: The Unsung Hero of Comfort

When we think HVAC, temperature often steals the spotlight. However, the humidity level in your home plays a pivotal role in overall comfort. Proper humidity control prevents mold growth, protects wooden fixtures, and even makes the air feel more comfortable at any temperature. Modern HVAC systems often integrate humidity sensors and controls to maintain optimal levels. Too dry? Your skin and sinuses suffer. Too damp? Hello, mold! A well-chosen HVAC ensures neither extreme trouble you, delivering that "just right" atmosphere. As you delve into HVAC options, prioritize systems that offer robust humidity management - it's the detail that makes all the difference.The Installation Process

Pre-installation Preparations:

Site Assessment: Laying the Groundwork for Success

Before even a single HVAC component lands at your doorstep, there's homework to be done: the site assessment. Think of it as a 'health check' for your home. It's not just about where the system will go, but how your home's architecture, insulation, and even its orientation can affect HVAC performance. Professionals will measure room sizes, check insulation quality, and identify any structural challenges. By understanding the unique characteristics and demands of your home, you ensure the HVAC system you choose fits like a glove - optimized for efficiency and effectiveness right from the get-go.

Ready, Set, Equip: The Gear-Up Guide to HVAC

Before diving head-first into an HVAC installation, it's imperative to have all the right tools and equipment on deck. Picture this: You're waist-deep in the process, and you realize you're missing that crucial wrench. Frustrating, right? To avoid such pitfalls, create a comprehensive checklist. From gauges and hammers to specialized HVAC equipment, ensure everything is within arm's reach. Maybe even do a mock 'run-through' to familiarize yourself with each tool's function. Remember, in the world of DIY, preparation isn't just half the battle—it's the entire game plan for smooth sailing.

Safety First: The Golden Rule of HVAC DIY

Before you dive into the world of HVAC, let's chat safety. It's not just about wearing gloves or protective eyewear (though they're vital!). It's a mindset. Always be aware of electrical components; a simple misstep can lead to unwanted shocks. Ventilate your workspace—those refrigerants aren't nose-friendly. And never, ever bypass safety protocols thinking "It'll be fine." Got kiddos or pets? Ensure they're at a safe distance. Every tool, no matter how small, should be handled with care. In the grand realm of DIY, it's not just about making your space comfortable; it's about keeping YOU safe while doing it.

Actual Installation:

System Placement:

Outdoor Unit Location: A Crucial Consideration

The location of an outdoor HVAC unit is not a decision to be made lightly. Proper airflow is paramount; the unit should be positioned in a location free from immediate obstructions like walls or large plants. Efficient airflow ensures the system operates at optimal capacity and reduces wear and tear. Noise is another concern. While these units are designed to operate quietly, they still produce sound. Consider the proximity to living spaces, especially bedrooms. Finally, service accessibility is essential. Technicians will need to perform regular maintenance, so the unit should be easily reachable. In summary, when determining the location, prioritize airflow, consider noise implications, and ensure easy access for servicing.

Indoor Unit Location: Centralized Efficiency Meets Aesthetics

When placing your indoor HVAC unit, the location can make all the difference. A centralized placement is vital for ensuring uniform air distribution throughout your living spaces. This means rooms are cooled or heated consistently, enhancing comfort. Beyond functionality, aesthetics matter. The unit should integrate seamlessly with your home's interior, not becoming an eyesore. Think about how it will fit with your decor. Lastly, consider accessibility. Regular maintenance is smoother when the unit is within easy reach, without having to move furniture or climb ladders. In essence, the perfect spot blends efficient air distribution with visual harmony and accessibility for upkeep.

Ductwork Installation (if applicable):

Design & Layout: Streamlined Ductwork for Maximum Efficiency

In the world of HVAC, the path of least resistance is key. When planning ductwork, it's essential to aim for a design that's both efficient and straightforward. Minimizing bends and constrictions can significantly enhance airflow and reduce energy consumption. Every bend can introduce resistance, making your system work harder and shortening its lifespan. By adopting a more linear design, you ensure smoother airflow, resulting in quicker room temperature adjustments and potential savings on your energy bills. Remember, a well-laid duct system isn't just about the air it carries but the energy it conserves.

Sealing & Insulation: A Critical Aspect of HVAC Efficiency

In the realm of HVAC systems, sealing and insulation play paramount roles. Proper sealing is essential to prevent the escape of conditioned air, ensuring that every bit of energy expended is utilized effectively. In tandem, insulation is crucial for maintaining the desired temperature, acting as a barrier against external temperature fluctuations. Both of these factors are vital not only for the optimal performance of your HVAC system but also for its longevity and your overall energy savings. A well-sealed and adequately insulated system ensures energy efficiency, translating to reduced utility bills and a consistent indoor environment.

Electrical Connections:

Safety Precautions: Electricity, Your Silent Adversary

Diving into any HVAC task? First, a safety pause. The electricity that powers your HVAC system is beneficial, yet it's silently perilous. Before delving into maintenance, installation, or checks, ensure the power is switched off. No compromises. It's not just about preventing a zap; it's about averting serious injuries or even fatal accidents. Familiarize yourself with the system's power switch, typically located near the unit or in your electrical panel. Secure the area, informing all household members of your work. Remember, when handling HVAC systems, your mantra should be: "Safety first, always!" Electrical precautions are non-negotiable in your DIY journey.

Wiring: The Veins of Your HVAC System

It's the lifeblood of your HVAC, carrying electricity to every vital component. Ensuring each connection is snug and steadfast is paramount. Loose wires? Not on your watch! Adhere religiously to provided guidelines—skipping a step isn't an option. Proper grounding is the unsung hero, grounding excess electricity and keeping everything safe. It's not just about function; it's about the safety of your home and its occupants. Every wire, every connection, every ground has its purpose. Treat them with respect, diligence, and the care they deserve in your DIY endeavors.

Refrigerant Handling:

Charging: The Right Juice for Your Cooling System

Think of your HVAC as a thirsty machine, and refrigerant as its preferred drink. Just like we wouldn't enjoy an over sweetened beverage or one too watery, your system craves the perfect refrigerant balance. Always confirm you're using the correct type—no mix and matches here! Using the right amount is equally pivotal. Overcharging or undercharging can spell trouble, affecting performance and lifespan. Equip yourself with the manufacturer's guidelines and stick to them religiously. Your system's efficiency and longevity hinge on this simple yet crucial step. Bottom line? Give it the right juice, and it'll serve you cool comfort in return.

Leak Testing: Sniffing Out the Sneaky Drips

Imagine throwing a pool party, but the pool has a secret leak. Not much fun, right? Similarly, any unnoticed leaks in your HVAC can leave you sweating, both from heat and unexpected repair costs. Enter the superhero of this tale: specialized leak-testing tools. These gadgets are designed to hunt down even the most elusive drips, ensuring your system remains airtight. By securing every nook and cranny, you maximize efficiency and prevent refrigerant loss. A routine leak check isn't just recommended—it's an essential step towards ensuring your cool companion works flawlessly.

System Testing:

Initial Start-up: A Critical Step

After all the installation steps, the initial start-up of your HVAC system is paramount. At this juncture, it's crucial to methodically monitor the system. Ensure that each component starts properly and operates without any issues. It's about more than just turning the system on; it's about verifying that all the work done up to this point culminates in a well-functioning HVAC. Listen for any unexpected sounds, check for system alerts, and be prepared to review any abnormalities. A successful start-up is a testament to a job well done.

Performance Check: The Final Confirmation

Once your HVAC system is up and running, the real test begins. The goal? Confirm it reaches your set temperature smoothly. Turn it on, let it operate, and observe. Is it quiet, free of any peculiar sounds or unsettling vibrations? This is your system's report card moment. If it meets the desired temperature and operates silently, give yourself a pat on the back; your HVAC installation is a success!

Final Touches:

Securing All Components: Double-Check for Peace of Mind

Post-installation, a thorough sweep is crucial. Examine every bolt, screw, and connection. Just as you wouldn't want a loose bolt in a car engine, the same goes for your HVAC. Secure fittings are the backbone of a seamless operation. Be meticulous; ensure everything is tightly fastened. Your system's longevity and efficiency might just hinge on this simple yet pivotal step.

Setting Up Controls & Thermostats: Empower & Educate

Once your HVAC system is in place, it's time to master its controls. Ensure thermostats are positioned for easy access and visibility. Most importantly, give homeowners a concise walkthrough. Familiarize them with settings, ensuring comfort and efficiency go hand in hand. Their confidence in using the system starts with your clear, authoritative demo.

Post-installation Recommendations:

Advice on Initial Operation: Starting Strong

Post-installation, it's crucial to operate your HVAC system judiciously. Begin with moderate temperature settings and observe how your home adjusts. Avoid extreme fluctuations initially, letting the system stabilize. This ensures longevity and optimized performance. Remember, patience now sets the tone for consistent comfort ahead.

Maintenance Schedules: Keep It Running Smoothly

After installing your HVAC, scheduling regular maintenance is vital. Aim for biannual check-ups—once before summer's heat and again prior to winter. This ensures peak performance and uncovers minor issues before they escalate. Timely maintenance not only extends your system's lifespan but guarantees efficient operation throughout the year.

Troubleshooting Common Issues: Be Your First Responder

Once your system's up and running, hiccups can still arise. Familiarize yourself with common problems like thermostat misreadings or unusual noises. Often, simple resets or filter changes do the trick. But, knowing when to call a professional is key. A quick reference guide can be invaluable, allowing you to address small glitches without a hitch.

CHAPTER 5

BASIC MAINTENANCE AND CARE

Routine Checks:

Your HVAC's Health Checkup

Just as we need regular check-ups for good health, your HVAC does too. Monthly inspections catch minor issues, preventing bigger, costly problems down the road. Seasonal assessments prep the system for temperature shifts, ensuring efficiency. Skipping these can lead to reduced lifespan and increased energy bills. Be proactive: mark your calendar and make your HVAC's health a priority.

The Art of Early Detection

Routine checks aren't just ticking off maintenance boxes; they're about foresight. By examining your HVAC regularly, you can spot early signs of wear or imminent issues. Catching these signs - like unusual noises, reduced airflow, or unexpected system restarts - early on can save hefty repair costs later. Remember, prevention is always more cost-effective than cure.

Cleaning and Replacing Components:

Air Filters

Breathe Easier by Staying on Top of Maintenance

Your system's efficiency hinges on air filter condition. For peak performance, standard fiberglass filters should be checked and likely replaced every month, especially during heavy usage seasons. Pleated filters offer a longer lifespan, typically 3 months. But if your home battles dust or pet hair, or if allergies are a concern, inspect more frequently. Remember, a clean filter doesn't just optimize airflow – it safeguards your indoor air quality. Adjust schedules based on your observations and specific household conditions, ensuring both system longevity and a healthier living space.

Mastering Maintenance and Making the Right Choices

When it comes to reusable filters, consistency is key. First, turn off the system. Then, remove the filter and gently shake off loose debris. For a deeper clean, use a soft brush or vacuum. If it's particularly grimy, a mild soap and warm water rinse does wonders; just ensure it's thoroughly dried before reinstalling. As for disposable filters, not all are created equal. Opt for pleated varieties with higher MERV ratings for superior filtration. Remember, while premium filters might cost more upfront, the protection they offer your system and the air you breathe is priceless.

Ducts:

The Pulse of Your Home's Airflow

Your ducts are the silent carriers of comfort in your home, but they too need attention. Ideally, duct inspections should happen every two years. Look out for signs of dust accumulation, as it can restrict airflow and reduce system efficiency. But there's a sneakier adversary: mold. Living in dark, moist environments, mold in your ducts can compromise air quality. If your home feels more humid, or if there's a musty odor, it's time for a professional inspection. Regular check-ups ensure your ducts remain clean and your air stays fresh.

Ensuring Clean Airways

Cleaning your ducts is paramount for optimal air quality. For minor dust, remove vent covers and vacuum inside using a hose attachment. Wipe down with a damp cloth and let it air dry. However, for extensive dust buildup or mold suspicions, it's a job for the pros. DIY methods can't always reach deep recesses, and improper cleaning can spread contaminants. Professional duct cleaning services use specialized equipment to ensure thoroughness. The rule of thumb: if it's more than surface grime, or if you've never had them cleaned, consider calling in the experts.

Condenser and Evaporator Coils:

Keeping it Cool and Clean

Your HVAC's efficiency hinges on its condenser and evaporator coils. These components directly impact energy consumption and overall system health. To ensure they function optimally, seasonal or annual cleanings are imperative. Typically, cleaning them before summer kicks in is recommended, ensuring your system runs smoothly during peak use. However, if your system is frequently used or located in a dusty environment, more frequent checks might be necessary. Regular maintenance not only boosts efficiency but also prolongs your HVAC's lifespan.

Cleaning Done Right

When cleaning these crucial coils, safety and precision are paramount. Start by turning off the power to avoid electrical hazards. For light dirt on the coils, a soft brush can do the trick. For deeper grime, use a no-rinse coil cleaner available at HVAC stores. Spray it on and let it foam and work its magic. Rinse if required. Ensure the area is well-ventilated when using chemicals. Never use high-pressure water; it can bend the fins. Always inspect the fins for damage and straighten with a fin comb if needed. Regular, gentle cleaning extends your HVAC's efficiency and lifespan.

Drain Lines and Pans:

Keeping It Clear and Clean

Your HVAC's drain lines and pans play a pivotal role in preventing water damage and mold growth. Given their importance, monthly checks are a must, especially during peak usage months. These components can easily become clogged with dirt and algae, leading to unwanted overflows. Regular inspections will help you catch any buildup early. By ensuring that the lines are clear and the pans are free of standing water, you not only protect your home but also maintain the optimal performance of your HVAC system. Remember, a little monthly attention can save significant repair costs down the line.

A Clear Path to Efficiency

For a smooth-functioning HVAC, drain lines and pans need to be obstruction-free. Begin with a visual inspection for any noticeable blockages. Use a wet-dry vacuum to suck out any obstructions from the drain line. For the pan, remove any standing water. A mixture of equal parts water and white vinegar can be poured down the line to naturally eliminate mold and algae growth. Always ensure the power to your HVAC system is off when performing these tasks. Regular maintenance can keep issues like water damage or system inefficiencies at bay. When in doubt, consult a professional.

Fan Blades and Blower Components:

Keeping the Air Flowing

Ensuring consistent airflow is key to your HVAC's efficiency, and the fan blades and blower components are the stars of this show. A semi-annual inspection is crucial. Over time, these components can accumulate dust and debris, which can reduce efficiency and potentially harm the system. Every six months, with the system powered off, visually inspect the blades for any dust buildup or signs of wear. A soft brush or cloth can often do the trick for cleaning. The blower, being a bit more intricate, may require a deeper dive. Checking belts for tightness and wear is also beneficial. If you spot any significant wear or damage, it might be time for a professional's touch.

The Clean Sweep Guide

Let's dive straight into the heart of your HVAC: the fan blades and blower components. Proper upkeep guarantees prolonged efficiency. Firstly, always disconnect power before any cleaning, safety first! For the fan blades, use a soft brush to gently sweep off any dust accumulation. Rotate the blades manually to ensure all sides are addressed. For the blower, a damp cloth can help wipe away stubborn debris. Always ensure it's completely dry before reconnecting power. Periodically, inspect the belts for signs of wear or loosening. Tighten if necessary. While cleaning, if you notice any anomalies or excessive wear, it might be time to seek professional help. Regular maintenance ensures a happier, longer-lasting HVAC system.

Seasonal Preparations:

Getting Set for the Summer Sizzle

Before summer's heat hits, ensuring your cooling system is up to the task is essential. Start by replacing or cleaning the air filters – a simple step that maximizes efficiency. Next, give the outdoor condenser unit some love; clear any debris and trim back foliage at least 2 feet for proper airflow. Inspect coolant lines for any insulation wear and replace if necessary. Run a test cycle early, listening for unusual noises and feeling for cool air. If your system struggles or doesn't cool efficiently, it may need refrigerant or professional servicing. Remember, a cool summer begins with proactive preparation!

Winter Warm-Up Wisdom

As winter's chill approaches, it's vital to ensure your heating system is ready to embrace the task. Start by swapping or cleaning the air filters—this simple action guarantees optimal warmth distribution. Ensure the outdoor unit is free from fallen leaves or snow, ensuring unrestricted airflow. Insulate any exposed pipes to prevent freezing. Test the heating system well before the first cold snap, tuning in for any irregular sounds and verifying it radiates heat effectively. If the system seems sluggish or fails to warm your space adequately, it might need a professional touch or more coolant. Remember, gearing up in advance ensures toasty winter nights!

Common Issues and Troubleshooting:

Your HVAC Detective Hat

Dive into the detective realm of HVAC troubleshooting! Common culprits? A non-responsive thermostat often indicates power issues—check the batteries or circuit breaker. No cold or hot air might mean a compressor or burner problem, or possibly low refrigerant levels. Unusual noises can signal loose components, while persistently running systems could point to a thermostat glitch. If you notice reduced airflow, filters might be clogged or there could be obstructions in the ducts. Before ringing up a technician, ensure all power switches are 'on', filters are clean, and circuit breakers aren't tripped. Remember, the more you know, the quicker the solve!

DIY Before Dialing

Before reaching for the phone, tackle these DIY troubleshooting champions. First, power-check! Ensure the system and thermostat are powered on and settings are correct. Next, peek at those filters. Dirty ones suffocate airflow—clean or replace them. Inspect circuit breakers; a tripped one might be the mischief-maker. Ensure vents aren't obstructed and the outdoor unit isn't covered in debris. Double-check the thermostat's batteries and settings. Lastly, reset the system; sometimes it just needs a quick reboot. If these steps don't resolve the hiccup, then it's time to call in the pros. Pat yourself on the back for being proactive!

CHAPTER 6
TROUBLESHOOTING COMMON PROBLEMS

Understanding Warning Signs:

HVAC's SOS Signals

Your HVAC system communicates – not with words, but with signs! First, the orchestra of noises: clunks, screeches, or rattles aren't part of its usual symphony. If it's sounding more like a rock band than a soothing lullaby, take note. A sudden dip in performance? That's its way of saying, "I'm not feeling well." Also, keep a vigilant eye on your energy bills. A sudden spike (without a logical reason like extreme weather) is its coded message about a lurking issue. Master the art of reading these signals, and you'll always be a step ahead in ensuring its health.

The Power of Regular Check-ups

Think of your HVAC like a trusty vehicle. Just as you wouldn't skip car servicing, your HVAC deserves that same diligence. Regular inspections are your proactive approach to nip budding issues in the bud. Even if everything seems fine on the surface, there might be underlying problems that only a closer look can reveal. These check-ups can spot wear and tear, parts that need replacement, or any other hidden culprits. By investing a little time now, you're avoiding potentially big, costly disruptions later. Your HVAC will thank you with optimal performance and a longer lifespan. Be wise, inspect regularly.

Issues Related to Temperature Inconsistencies:

Thermostat Issues:

Decoding the Telltale Signs

The thermostat is your HVAC's brain, guiding it on when to kick in and chill out. However, even brains hiccup. If your HVAC suddenly plays the rebel—refusing to start, halting prematurely, or running an unending marathon—it might be crying out, "Check the thermostat!" These symptoms often suggest a miscommunication between your device and the system. Just like ensuring your phone and charger are compatible, your HVAC and thermostat must be in sync. Understanding these signs can save you from larger issues down the line. Be observant, and let your system whisper its needs to you.

Unearthing the Root Causes

When your thermostat acts up, don't sweat it—most issues have logical explanations. First, an innocent oversight like incorrect settings can be the culprit. It's akin to your alarm not ringing because it was set for PM instead of AM! Next, sensors inside can falter over time, losing their sharpness like an old pair of spectacles. And lastly, wiring problems can arise, much like an interrupted phone call with weak reception. Knowing these potential causes empowers you to diagnose quicker and act smarter. Stay cool, and keep those HVAC tunes harmonious!

Swift Solutions at Your Fingertips

Facing thermostat troubles? Before you hit the panic button, try these easy-peasy solutions. Start by giving your thermostat a fresh start—like rebooting a computer—with a simple reset. Next, take a moment to verify its program settings. It might just be like when you accidentally set your alarm for Saturday instead of Monday. Finally, check if it's playing the right seasonal tune: ensure it's set to 'heat' for winter warmth or 'cool' for summer breezes. Often, the simplest checks save the day, keeping your indoor vibes just right!

Blocked or Leaky Ducts: Navigating the Hidden Passages of Your HVAC

So, you've noticed that your living room feels like the Sahara Desert while your bedroom seems more suited for penguins? Uneven heating or cooling is a common household woe, but worry not! It's often due to pesky blocked or leaky ducts.

Symptoms: Where's the Heat (or Cool)? Your HVAC system is like the heart of your home, pumping warm or cool air through a network of ducts. When everything is smooth, each room gets its fair share of treated air. But if you're finding temperature inconsistencies, with some rooms feeling too hot or cold, you might be dealing with the sneaky culprits of blocked or leaky ducts.

Potential Causes: What's Clogging the Arteries?

Obstructions in the Ducts: Dust, debris, or even critters could be having a party in your ducts. Over time, they accumulate, especially if the system hasn't been cleaned for an extended period.

Holes or Disconnects: Over time, ducts can suffer wear and tear. Shifting foundations, frequent temperature changes, or simply age can lead to holes or sections of duct becoming disconnected. This is like having a hole in a water pipe, where the air seeps out, never reaching its intended destination.

Quick Fixes: DIY to the Rescue

Before calling in the HVAC cavalry (which, honestly, might be needed for some severe cases), here are a few steps to help you navigate and perhaps resolve the issue:

Visual Inspection: Start by visually inspecting the portions of the ductwork you can access—basements, attics, and crawlspaces are good starting points. Look for any glaring disconnects. Sometimes a section can simply slip apart, especially if it wasn't secured properly in the first place.

Feeling the Airflow: Turn on your HVAC system and place your hand near any suspected leaks. If you feel air blowing out or even a slight draft, you've identified a spot that needs attention.

Clearing Obstructions: If you find a blockage, and it's within reach, put on a pair of gloves and gently remove it. Sometimes it's as simple as accumulated dust or debris; other times, it might be something more substantial, like insulation that has slipped into the ducts.

Duct Tape to the Rescue: For those minor leaks you've identified, duct tape can be a temporary solution. Ensure the surface around the leak is clean and dry. Apply the tape generously, pressing firmly to ensure it adheres properly. It's worth noting that while duct tape is a quick fix, it's not a permanent solution. Over time, the adhesive can deteriorate, especially with temperature fluctuations. For a longer-lasting seal, consider using mastic sealant or a metallic foil tape.

Professional Cleaning: If you suspect obstructions but can't find them in the accessible duct sections, it may be worth considering a professional duct cleaning. They'll have the tools and expertise to thoroughly clean the entire system, ensuring smooth airflow throughout.

Keep 'Em Clean: Going forward, consider periodic duct inspections as part of your home maintenance routine. Cleaning or replacing your HVAC filters regularly can also reduce the amount of debris that ends up in the ducts.

Closing Thoughts: While these DIY fixes can handle minor duct issues, remember that significant leaks or damage will require professional attention. After all, your duct system is a vital part of your home's comfort, and ensuring it's in tip-top shape will not only keep each room cozy but can also save you on energy bills.

Remember, every home has its quirks, and uneven heating or cooling is just one of the adventures in homeownership. With a bit of proactive care, attention, and sometimes a roll of duct tape, you'll be well on your way to mastering the hidden passages of your HVAC!

Dirty or Clogged Filters: The Silent Troublemakers in Your HVAC System

Ah, air filters! These seemingly innocuous sheets, often out of sight and out of mind, play a pivotal role in ensuring the smooth operation of your HVAC system. Like a diligent goalkeeper stopping unwanted invaders, they trap dirt, debris, and other microscopic undesirables. However, even the best goalkeepers need a break now and then, and that's what this chapter's all about.

Symptoms: Is My Filter on Strike?

Just as a straw becomes hard to sip from when it's pinched, a clogged air filter can restrict the flow of air in your HVAC system. But how do you recognize the signs?

Reduced Airflow: If you place your hand near a vent and the flow of air feels weaker than usual, this is a telltale sign. Remember those refreshing gusts during summer or the warm embrace in winter? If those seem less intense, it's time to check the filter.

HVAC Working Overtime: Hear your HVAC system making more noise than usual? Or do you feel it's running longer to achieve the desired temperature? A clogged filter might make the system work harder to circulate air.

Potential Causes: What's Choking My Filter?

The main job of an air filter is to ensure the air circulating in your home is clean. But with great responsibility comes the accumulation of unwanted guests:

Accumulated Dirt: Everyday dirt and dust from our living environment tend to be the usual culprits. This is the day-to-day stuff – think about dust from open windows, pet dander, or just the general wear and tear of daily life.

Debris: Think larger particles here. If there's construction nearby or perhaps you've recently undertaken a DIY home improvement project, this can introduce larger debris into the air.

Prolonged Time Since Last Replacement: Filters don't last forever. Over time, the accumulated particles will block the filter. If you can't remember the last time you checked or replaced the filter, it's probably overdue.

Quick Fixes: Breathe Easy Again

The beauty of addressing dirty or clogged filters is that the fixes are usually quick, easy, and inexpensive. Here's how to restore that clean, free-flowing air:

Visual Inspection: Pull out your filter (make sure to turn off your HVAC system first) and hold it up to the light. If it looks like it's wearing a thick winter coat of dust and dirt, then it's definitely time for a clean or replacement.

Cleaning Reusable Filters: If you've invested in a reusable filter, here's how to spruce it up:

- Start with a gentle vacuum to remove the loose particles.
- Rinse under cold water. Avoid using any cleaning agents as they might damage the filter or introduce unwanted chemicals into your air.
- Ensure the filter is completely dry before reinserting. A damp filter can become a breeding ground for mold.

Replacing Disposable Filters: For those with disposable filters, it's a straightforward swap:

- Purchase the right size and type. If unsure, the specifications are usually printed on the edge of your current filter.
- Insert the new filter, ensuring it's oriented correctly (there's usually an arrow indicating airflow direction).

Frequency Is Key: A general rule of thumb is to check your filter monthly, especially during high-usage months. Depending on the cleanliness of the surrounding environment and filter type, replacements or cleaning can be done every 2-6 months. If you have pets or are in an area with high pollen counts, consider checking more frequently.

Opt for Quality: While it might be tempting to save a few bucks on cheaper filters, investing in a good-quality filter can pay off in the long run. Not only do they last longer, but they also filter out smaller particles, ensuring better air quality.

The health of your air filter is synonymous with the health of your home environment. An effective, clean filter ensures that the air you breathe is of the highest quality and that your HVAC system runs efficiently without undue strain. Regularly checking, cleaning, or replacing your filter is a simple yet critical DIY task that can have significant benefits for both your health and your wallet. So next time your HVAC seems to be working a tad too hard, remember to give its unsung hero – the air filter – a little TLC!

Malfunctioning Components: The Nuts and Bolts Behind the Breeze

We often marvel at the wonder of modern HVAC systems. A simple twist of a dial or a press of a button, and voila! Our rooms transform into a summer haven or a winter retreat. But behind that magic are a multitude of components working in perfect harmony. What happens when one of these key players strikes a wrong chord? Let's dive into the world of malfunctioning components.

Symptoms: When the Air Just Isn't Right

Ever set the thermostat and patiently waited for that refreshing gust of air, only to be greeted by...well, room temperature blandness? It's like eagerly expecting a symphony and getting a recorder solo.

Unmet Expectations: The most straightforward symptom is when your HVAC system isn't producing air at the temperature you've set. It's not too hot, not too cold, just lukewarm.

Potential Causes: The Culprits Behind the Dissonance

Understanding the machinery behind the thermostat can help us grasp why sometimes, things just don't heat up (or cool down) as they should:

Faulty Compressor: Consider the compressor the heart of your HVAC system, particularly for the cooling function. When the compressor isn't doing its job, the entire cooling process can falter. A faulty compressor may not be compressing the refrigerant properly, leading to inadequate cooling.

Malfunctioning Evaporator Coils: The evaporator coils are where the real magic happens, turning refrigerant from liquid to gas and removing heat from the air. If these coils are malfunctioning, you'll feel it – or more accurately, you won't feel that cooling effect.

Problematic Condenser Coils: These are the evaporator coils' counterpart, responsible for releasing accumulated heat outside your home. If they're not working correctly, the system struggles to cool down, no matter how low you set that thermostat.

Quick Fixes: What Can a DIY Enthusiast Do?

Now, I must preface this section with a word of caution. Many issues related to malfunctioning components necessitate a professional's expertise. However, for the hands-on folks among us, there are a few checks you can conduct before ringing up your HVAC technician:

Visual Inspection: Ensure the unit is off and unplugged. For the outdoor unit (usually housing the compressor and condenser coils), check for visible blockages, like leaves or other debris. A build-up of dirt or debris can affect airflow and overall system performance.

Clean the Coils: While you've got those units open, take a look at the coils. Do they look dirty or obstructed? Gently clean them using a soft brush or cloth. For a deeper clean, a mix of water and mild detergent can work, but make sure it's thoroughly rinsed and dried. Remember, while evaporator coils might be accessible in some indoor units, they can be trickier to reach in others, so proceed with caution.

Check for Ice: Sometimes, especially with evaporator coils, a malfunction can lead to ice formation. If you spot this, it's a clear sign something's amiss. Turn off your system and let the ice melt. While this won't solve the root problem, it's a step in the right direction.

Inspect the Compressor: With the power off, take a look at the compressor. Any visible damage? Does it seem to be vibrating oddly when in operation? While you might not be able to fix a compressor issue yourself, identifying the problem can be a great help when explaining the situation to a professional.

Thermostat Check: It might sound obvious, but sometimes, the issue is as simple as a malfunctioning thermostat. Ensure it's set correctly and has working batteries.

When to Call in the Calvary

After your checks, if the system still isn't performing up to par, it's time to call in an HVAC technician. The complexities of components like compressors, coils, and the intricate dance of refrigerants make professional intervention not just recommended, but often essential. They'll have the tools, expertise, and replacement parts to get your system back in tip-top shape.

The world inside our HVAC systems is one of intricate coordination and precise functionality. When one component is out of tune, the whole symphony can feel off. By understanding the key players, recognizing the signs of malfunction, and performing basic DIY checks, you arm yourself with knowledge. And when all else fails, remember: there's no shame in calling for an encore with the help of an HVAC maestro.

Electrical and Wiring Issues: Sparks, Shorts, and Safe Solutions

Ah, electricity. The invisible powerhouse behind every appliance and gadget in our homes. It's the lifeblood that keeps our HVAC humming and homes cozy. But when things go awry in the electrical realm, it's a different story. These issues require a blend of detective work and a heaping dose of caution. Let's embark on a journey through the maze of wires and circuits!

Identifying the Telltale Signs: Electricity's Silent Whispers

When electrical problems begin to manifest in your HVAC system, they often leave subtle clues behind. Recognizing these early signs is pivotal for safety and the longevity of your system:

Unusual Sounds: Hissing, buzzing, or sizzling noises can be a sign of electrical issues. If your HVAC unit sounds more like a busy kitchen than a standard appliance, you might be dealing with a problem.

Tripping Circuit Breakers: If your system is constantly tripping the breaker, it's not just being quirky; it's signaling an overload or short circuit.

Flickering or Dimming Lights: When the HVAC kicks in, do your home's lights dim or flicker? This can suggest an electrical imbalance or inadequate power supply.

Burnt Odors: The unmistakable scent of burnt plastic or wiring is a red flag. It's a clear indication of overheating components or frayed wires.

Non-Responsive System: Pushing buttons with no response? If your HVAC refuses to start up, electrical connectivity issues might be the culprit.

Safety First: The Non-Negotiables of Dealing with Electrical Issues

Messing with electricity isn't like any other DIY project. The stakes are high, and the margin for error is minimal. Here are some golden rules:

Power Down: Before investigating, always turn off the power to your HVAC system at the breaker box. This simple step can prevent dangerous shocks.

Hands Off Exposed Wires: If you spot any frayed or exposed wires, don't touch them, even if the power is off. Damaged wires can be unpredictable.

Avoid Water: Always ensure your hands and the surrounding area are dry. Water and electricity are a dangerous duo.

Use Insulated Tools: If you decide to do a bit of investigative work, only use tools with rubber or insulated grips. These are designed to protect you from accidental electrical contact.

Stay Grounded: Literally. Wear rubber-soled shoes to provide an additional layer of protection against potential shocks.

Calling in the Pros: Knowing When It's Time to Wave the White Flag

Electrical issues can be complex and, frankly, intimidating. While it's admirable to want to solve problems on your own, there are moments when it's best to defer to the experts:

Persistent Issues: If you've tried basic troubleshooting and the problem keeps returning, it's time to call a professional.

Complex Wiring Systems: Modern HVAC units have intricate wiring systems. If you're unsure about any aspect, it's better to step back and let an expert handle it.

Safety Concerns: Feel out of your depth? Or concerned about potential hazards? Trust your gut. There's no DIY pride worth compromising safety.

The dance of electrons behind your walls is a delicate and powerful ballet. When your HVAC system starts showing signs of electrical issues, it's crucial to approach the situation with a mix of curiosity and caution. Understand the symptoms, always prioritize safety, and recognize when it's time to hand over the reins to a professional. Remember, in the world of electricity, respect, awareness, and a bit of humility go a long way.

Water and Refrigerant Leaks: Navigating Drips, Drops, and Potential Disasters

Alright, DIY aficionados, let's talk leaks! No, we're not referencing that pesky faucet in the bathroom or the dripping kitchen sink. We're diving into the sometimes-murky waters of HVAC leaks—specifically, water and refrigerant leaks. Both can be a cause for concern but armed with the right knowledge, you can tackle them head-on.

Spotting the Leaky Culprits: From Puddles to Pools

Leaks love leaving breadcrumbs (or rather, water droplets). Here's what you need to be on the lookout for:

Pooling Water: If there's a mini-lake forming around your HVAC unit, it's a blatant sign. But even smaller amounts of collected water should raise eyebrows.

Frosty Coils: Ice on the evaporator coils? This could be a sign of a refrigerant leak, causing the coils to freeze.

Musty Odors: Persistent, unexplained musty smells can be a sign of mold growth due to ongoing water leakage.

Hissing Sounds: A refrigerant leak might not always be visible, but sometimes, it's audible. A gentle hissing from the unit can hint at escaping refrigerant.

Waterfall Origins: Why's My HVAC Leaking Anyway?

Understanding the *why* behind a leak is half the battle. Here are some potential culprits:

Clogged Drain Pan or Line: Dust, dirt, or mold can block the drain line, causing water to overflow from the drain pan.

Frozen Evaporator Coils: Low refrigerant levels or inadequate airflow can freeze the coils, leading to a water overflow as they defrost.

Cracked Drain Pan: Time and wear can degrade the drain pan, leading to water leaks.

Faulty Seals: Refrigerant leaks often occur when seals or joints in the refrigeration system weaken over time.

A DIYer's Response to Leaks: From Detection to Action

When you find a leak, don't panic! Instead, spring into action with these steps:

Switch Off: Turn off your HVAC system. Continuing to run a leaking unit can compound the problem and lead to more significant damage.

Contain the Leak: Place absorbent towels or buckets around the leaking area to prevent water damage. This is especially crucial for water leaks which can damage floors or furnishings.

Inspect Visually: Check the drain pan for cracks or damage. If there's a blockage in the drain line, try to clear it using a wet/dry vacuum.

Refrigerant Caution: If you suspect a refrigerant leak, it's essential to proceed with caution. Refrigerant isn't just harmful to the environment; it can pose health risks if inhaled.

Call a Pro for Refrigerant: While some issues like a blocked drain line can be DIY-ed, a refrigerant leak needs a professional touch. They'll need to locate the leak, fix it, and then refill your system with the right amount of refrigerant.

Stay Dry and Breathe Easy

Leaks might feel like a homeowner's nightmare, but with the right approach, they're more of a hiccup. Recognize the signs early, pinpoint potential causes, and either roll up those DIY sleeves or recognize when to call in the experts. Here's to dry floors, optimal HVAC performance, and the confidence to navigate those unexpected drips and drops!

CHAPTER 7

ENERGY EFFICIENCY AND SUSTAINABILITY

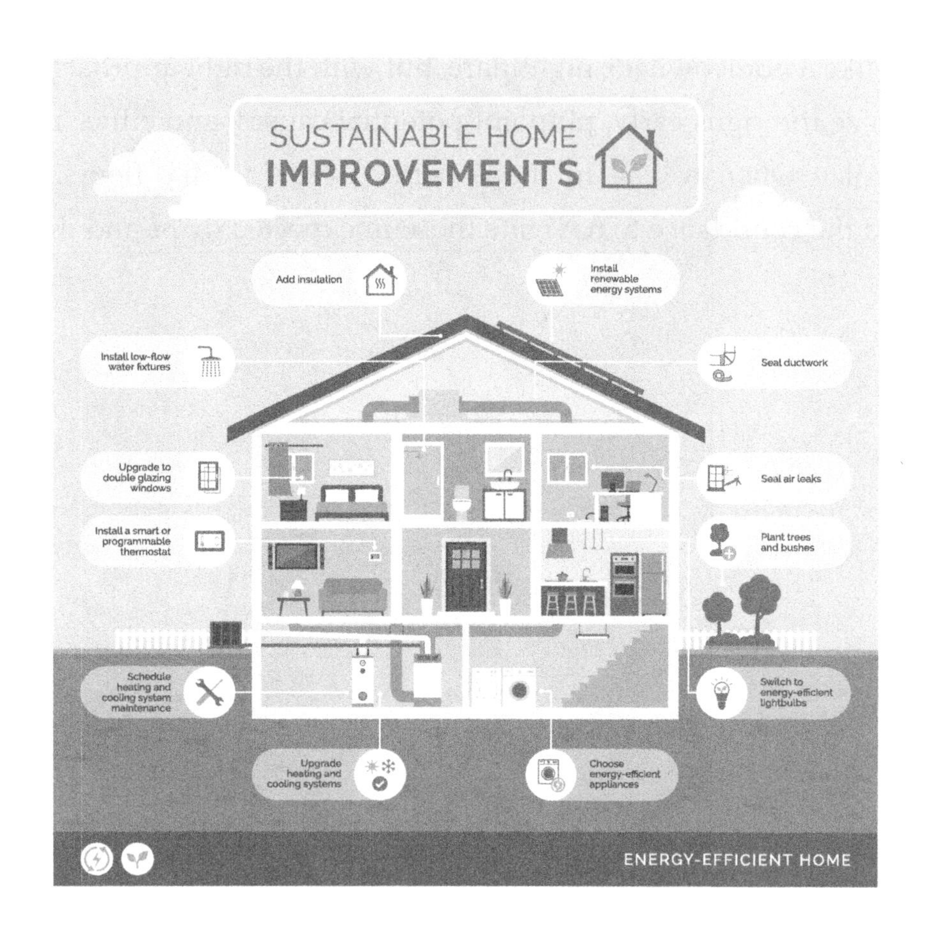

The Relationship Between HVAC Systems and Energy Consumption: More than Just Hot Air

HVACs: The Energy-Guzzling Giants

Your HVAC system could be the *largest energy consumer* in your home. Shocking, isn't it? Here's a breakdown:

Power Play: While our modern comforts like fridges, ovens, and washing machines use energy, HVAC systems outpace them all, often accounting for up to half of a home's total energy usage. They're working tirelessly to keep you cozy in winter and chill in summer.

Inefficiencies Multiply: An inefficient HVAC system doesn't just work sub-optimally. It works harder, runs longer, and consumes more energy doing the same job as a newer, more efficient model.

More than Just Electricity: It's not just about the power bill. Some HVAC systems, especially older models, use fuels like oil or natural gas. These not only cost more but can also have efficiency losses, translating to wasted energy (and money).

Eco-Footprint: What Your Old HVAC Might be Hiding

Okay, so we've established that HVACs can be quite the energy hog. But what does that mean for our dear planet?

Carbon Emissions: Energy consumption equals carbon emissions, especially if your electricity source isn't green. An inefficient HVAC system contributes more CO2 to the atmosphere, accelerating global warming. Not the kind of legacy we want, right?

Refrigerants – The Silent Polluters: Older HVAC models often used refrigerants like R-22, known for depleting the ozone layer. While many of these are being phased out, if you've got a grandpa HVAC, it might still be using these environmentally harmful chemicals.

Wasted Resources: Manufacturing, transporting, and installing HVAC systems require resources. An inefficient system that needs replacing more frequently means more materials used and more waste generated.

A Glimmer of Hope: What Can We Do?

Now that we've outlined the potential issues, it's time for the silver lining. There's a lot homeowners can do:

Regular Maintenance: Just like you wouldn't run a marathon in old, worn-out shoes, don't let your HVAC work with dirty filters or clogged ducts. Regular check-ups can improve efficiency.

Upgrade Time: If your HVAC system is older than that shirt you've been meaning to throw out (you know the one), consider upgrading. Modern systems come with better efficiency ratings and use environmentally friendly refrigerants.

Seal the Deal: Ensuring that your home is well-insulated means your HVAC doesn't have to work as hard. Seal those windows, insulate those walls, and enjoy a more comfortable home and a happier planet.Wrapping It Up: Breathe Easy & Tread Lightly

Your HVAC system, with all its hums and buzzes, plays a silent role in both your home's energy consumption and its environmental impact. But with knowledge comes power. Recognize its potential pitfalls, and take action to make it a green champion rather than an eco-villain.

Advancements in HVAC Technologies for Improved Efficiency:

Variable Speed Technology: More Than Just a Fancy Name!

Ever wondered if there's a smarter way to manage your home's heating and cooling, one that doesn't involve manually fiddling with the thermostat every hour? Enter: Variable Speed Technology. This chapter will unravel the intricate dance of how this technology precisely adjusts to your home's demands and why it might just be the best thing since sliced bread for HVAC enthusiasts.

What's In a Name? Understanding Variable Speed Technology

Now, before you drift off into tech-jargon-induced slumber, let's demystify this term. The 'speed' here refers to the blower motor inside your HVAC system. Traditional systems typically operate at two speeds: 'on' and 'off'. Picture it like a car that only knows two speeds: full throttle or complete stop. Not very efficient, right?

Variable Speed Technology, on the other hand, can be likened to a car that adjusts its speed based on the road conditions. Instead of just being "on" or "off", these systems can modulate their output, operating anywhere from 25% to 100% capacity, adapting in real-time to the cooling and heating demands of your home.

Riding the Energy Wave: The Immense Benefits of Variable Speed

Alright, now that you have a basic grasp of how this tech operates, let's delve into why you should care. Spoiler alert: It's not just about being fancy!

Energy Savings Galore:

The Modulation Magic: By not always running at full tilt, variable speed systems can save considerable energy. Imagine the gas savings if your car engine idled smoothly in traffic rather than revving at max capacity. The same principle applies to your HVAC system.

Peak Efficiency: These systems reach their desired temperature quickly and maintain it efficiently, ensuring no energy is wasted.

Your Comfort, Elevated:

No More Temperature Swings: We've all experienced those unpleasant temperature spikes or drops when an old HVAC kicks in or shuts off. Variable speed systems provide consistent airflow, eliminating these drastic temperature swings.

Whisper-Quiet Operations: As a bonus, these systems tend to be much quieter, given that they're not always running at full blast. No more being startled awake by the sudden roar of your heating kicking in!

Long Live Your HVAC!:

Reduced Wear and Tear: Traditional systems undergo a lot of stress from the frequent stop-start cycle. By operating continuously at a lower capacity, variable speed technology reduces the strain on the system components.

Fewer Breakdowns: Consistent operations mean fewer breakdowns and potentially longer life for your HVAC system. And we all love a device that doesn't give up on us, right?

In a Nutshell: Speed Doesn't Always Mean Rushing

To bring this chapter to a snug close, variable speed technology isn't about rushing but adjusting. It's the difference between a one-size-fits-all approach and a tailored suit. Your home deserves that tailored fit, one that brings comfort, efficiency, and longevity.

In the vast ocean of HVAC technologies, variable speed stands out as a beacon of smart energy use. For those considering an upgrade or a new installation, this is undoubtedly a feature worth considering. Not just for the energy savings (which, let's face it, are substantial!) but for the sheer comfort and peace of mind it brings.

Your home is your sanctuary, and with variable speed technology, it just got a lot cozier, quieter, and eco-friendlier. So the next time you're pondering about heating and cooling solutions, remember: it's not always about speed, but about the wisdom of modulation!

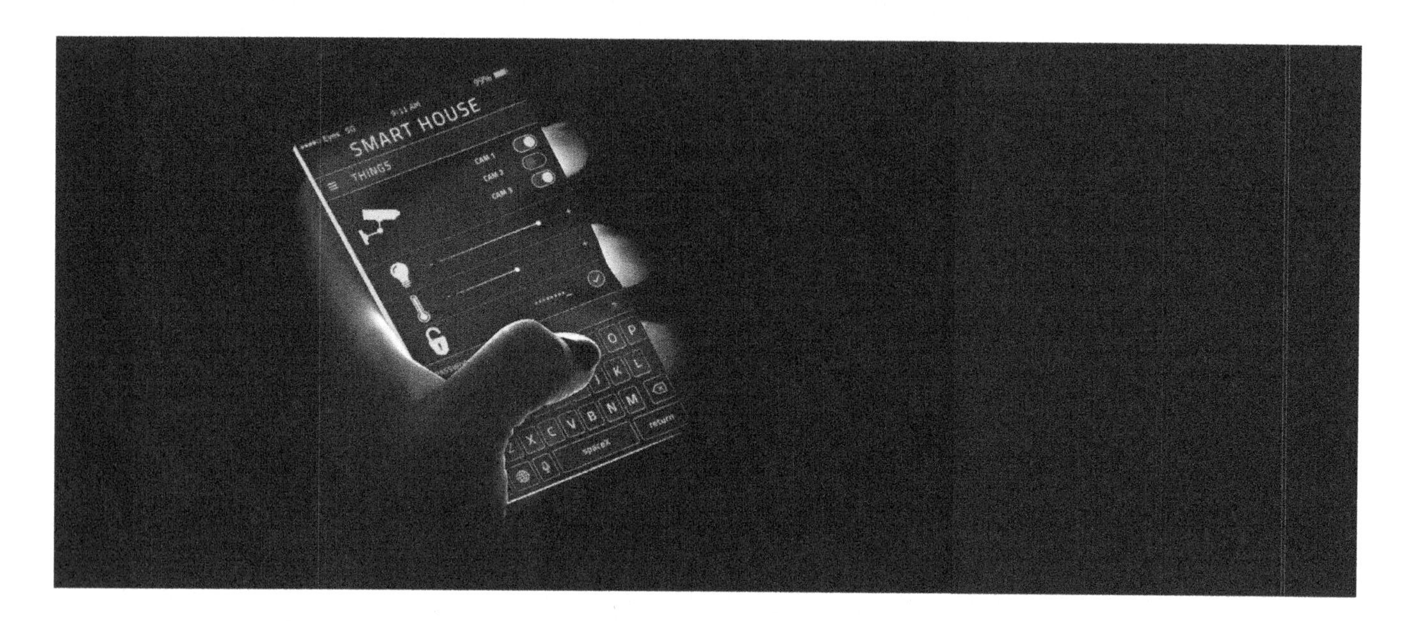

Smart Thermostats and IoT Integration: Stepping Into The Future of Home Comfort

Understanding The Marvel: Smart Thermostats

If you've ever wished your home could 'get' you, understand your unique quirks and preferences, then smart thermostats are your genie granting that wish. These aren't your grandparents' thermostats; they're the James Bonds of the HVAC world—sleek, smart, and ultra-efficient.

At their core, smart thermostats do what any thermostat does—regulate your home's temperature. But then, they sprinkle a bit of magic. They *learn* from your behaviors and preferences, adjust heating or cooling as you come and go, and even allow you to control them from anywhere in the world with an internet connection. Lost in the Swiss Alps but want your home toasty when you return? There's an app for that.

IoT Integration: Making the Connection

Before we move further, let's demystify IoT. Standing for 'Internet of Things', IoT is like an invisible thread connecting all your smart devices, letting them talk and work together. So, when your smart thermostat is integrated with other smart home devices via IoT, magic unfolds. Your lights, security system, blinds, and even your coffee machine can sync with your HVAC rhythms.

Reaping The Rewards: Why Smart and Connected Is The New Cool

Precision at Your Fingertips:

- *Tailored Comfort*: These thermostats don't just react; they predict. Over time, they learn your routines and adjust accordingly, ensuring your home's temperature is always to your liking.
- *Remote Control*: Gone are the days of dashing home in panic, recalling you left the AC on. With your smartphone, you can adjust settings from anywhere.

Optimized Energy Consumption:

- *Energy Reports*: Knowledge is power. Many smart thermostats provide detailed energy usage reports, helping you understand and adjust your consumption habits.
- *Eco-Friendly Modes*: These devices often come with eco-friendly settings, ensuring that while you enjoy optimal comfort, Mother Earth breathes a tad easier.

Synchronized Smart Living:

- *Integrated Routines*: With IoT integration, your morning routine might look like this: your thermostat slightly raises the temperature, your blinds open, and your coffee machine starts brewing—all seamlessly synchronized.
- *Enhanced Security*: Going on vacation? Your thermostat can sync with your security system. If there's unexpected movement, the system can make it seem like you're home by adjusting the lights and temperature.

Cost Savings:

- *Efficient Operations*: By learning your routines and automatically adjusting settings, these thermostats ensure no energy is wasted. This means more money stays in your pocket.
- *Informed Decisions*: Those energy reports we talked about earlier? They aren't just fun graphs. They can show you when and where you're consuming the most energy, allowing for strategic adjustments.

Future-Proofing Your Home:

- *Regular Updates*: Like any smart device, these thermostats receive regular software updates, ensuring they're always equipped with the latest features and security enhancements.
- *Compatibility*: As the world of smart home devices expands, having a connected system ensures that adding new devices to your home ecosystem is a breeze.

Bringing It All Together: A Home That Thinks

To wrap this chapter snugly, we're not just talking about a piece of tech. We're envisioning a home environment that evolves with you, one that knows when to envelop you in warmth or cool you down, even before you realize it yourself.

Smart thermostats and IoT integration are not fleeting tech trends; they are the future of sustainable, comfortable, and efficient living. By embracing this tech, you're not only stepping into the future—you're crafting it, one precise temperature setting at a time!

So, as you ponder the future of your living space, remember: homes can think, adapt, and care. The future isn't just smart; it's brilliant.

Zoning Systems: Crafting Custom Comfort Room by Room

Zoning Systems Demystified: One Home, Multiple Climates

Let's begin our journey by understanding the very essence of a zoning system. Picture your home as a big theatre, and each room is its own unique act with different needs. The living room, bustling with activity, may require a different temperature than the cozy, sunlit bedroom upstairs. With a conventional HVAC system, you set one temperature, and it's like making every actor wear the same costume regardless of their role! Zoning systems, however, let each 'actor' (or room) shine in their own distinct outfit (or temperature).

Here's how it works. The system divides your home into different zones, each controlled by its own thermostat. Dampers in the ductwork control and direct conditioned air only where it's needed, giving each zone the ability to have its own unique temperature.

Savoring the Zoned Life: Benefits of Going Bespoke

Tailored Comfort:

- *Personal Spaces*: Each family member can enjoy their preferred temperature in their personal space. The artist might need a cooler home office while the granny prefers a warmer living room.
- *Adapting to Home Layout*: Homes aren't uniform. Rooms with large windows might become sunbathed paradises by noon. Zoning caters to these nuances, ensuring each space feels just right.

Efficiency and Energy Savings:

- *Wise Energy Use*: Think of it like this - why have all the lights in the house on when you're just in the living room? Similarly, why condition areas of the home that are unoccupied? Zoning ensures you're only using energy where it's truly needed.
- *Reduced Strain on the System*: By not forcing your HVAC to condition the entire home uniformly, you reduce its workload. This could translate to a longer system lifespan and fewer maintenance hiccups.

Financial Frugality:

- *Sliced Energy Bills*: By ensuring conditioned air is only directed where necessary, you could see a reduction in your energy bills. It's like only filling the gas tank as much as you need, rather than always filling it up!
- *Increased Property Value*: A zoned home is an attractive prospect for future buyers who might be eyeing energy efficiency and customized comfort.

Enhanced Flexibility:

- *Future Home Projects*: Converting the garage into a gym? Or maybe the attic into a reading nook? With a zoning system in place, integrating these new spaces into the home's comfort landscape becomes seamless.

- *Seasonal Adjustments*: As the seasons dance by, certain areas of your home might have different conditioning needs. The south-facing summer room might need more cooling during a sultry July afternoon. Zoning allows for such dynamic adjustments.

A Few Considerations

While zoning sounds like a dream, and in many ways it is, it's crucial to ensure it's the right fit. Here are some things to keep in mind:

- **Home Assessment**: Before diving into zoning, get a professional to assess your home's layout, occupancy patterns, and existing HVAC capabilities. This ensures the zones are established logically.
- **Installation and Maintenance**: Zoning systems, with their dampers and multiple thermostats, might require a more intricate installation and maintenance approach. Always work with trusted professionals.
- **Cost vs. Savings**: Yes, there's an upfront cost. However, remember to weigh this against potential long-term energy savings and the unparalleled comfort it offers.

As we wrap up this chapter, imagine the joy of walking from one room to another, each greeting you with its own climate embrace. That's the beauty of zoning. It's the realization that comfort isn't one-size-fits-all. In our homes, just as in life, sometimes we need a touch of bespoke.

So, the next time you're in a room that's too chilly or a space that feels like a mini-Sahara, remember: your home can and should dance to your rhythm. And with zoning systems, every room gets its own unique dance move!

Geothermal Heating and Cooling: Tapping Earth's Temperate Treasure

Delving Deep: The Mechanics of Geothermal Systems

Imagine, for a moment, burrowing beneath the ground. As we journey deeper, we find an area where surface weather changes don't reach, and temperatures remain remarkably stable. This underground realm is our ally in geothermal heating and cooling.

A geothermal system typically involves a "loop" of pipes buried in this underground world. In winters, these loops absorb Earth's natural warmth, amplify it using a heat pump, and then circulate this heated air throughout your home. Come summer, the process reverses: the system extracts excess heat from your home, releases it into the cooler ground, and sends chilled air back. Essentially, it's leveraging Earth's stable temperatures as a natural HVAC mechanism!

The Bounty of Benefits: Embracing Geothermal

Peak Efficiency:

- *Stable Source*: The ground temperatures remain steady, enabling geothermal systems to function more efficiently than traditional HVAC units that combat fluctuating outdoor temperatures.
- *Energy Consumption*: Geothermal units can use up to 50% less electricity compared to their traditional counterparts, making them a boon for energy-conscious homeowners.

Long-Term Savings:

- *Reduced Operating Expenses*: Though the initial installation might pinch the pocket, the subsequent drop in monthly energy bills will bring a smile to your face. The initial investment often pays for itself within a few years.
- *Durability*: A hallmark of geothermal systems is their longevity. With indoor components typically lasting about 25 years and the ground loopwork functioning for up to 50 years, you're setting yourself up for decades of comfort.

Eco-Friendly Operations:

- *Lower Carbon Footprint*: Reduced electricity consumption translates to fewer greenhouse gas emissions. With geothermal, your home becomes an environmental steward.
- *Conservation of Resources*: Since geothermal systems don't rely on fossil fuels, they sidestep the pollution and environmental degradation linked with their burning.

Quiet Functioning and Enhanced Safety:

- *Noiseless Operation*: Unlike traditional systems, geothermal units operate almost silently, offering comfort without the accompanying soundtrack.
- *Safety and Landscape Benefits*: Lacking external units, there's no need to design your landscape around them. Additionally, with no combustion involved, risks related to open flames or potential carbon monoxide leaks are eliminated.

Points to Consider

Before you set your heart on a geothermal system:

- **Initial Costs**: The excavation and installation of underground loops can be investment-heavy. However, given the long-term savings, many find this a worthy investment.
- **Assessing Property**: It's vital to evaluate if your property is suitable for geothermal. Factors like plot size, soil composition, and underground obstacles can influence installation feasibility.
- **Rely on the Experts**: Geothermal installation and maintenance are not for the average DIYer. Ensure you engage with experienced professionals for optimal results.

Embracing Earth's Steady Embrace

As you ponder on a geothermal system for your property, reflect on the harmonious synergy it represents: a blend of innovative engineering and Earth's natural stability. It's not just about heating or cooling; it's about embracing an eco-friendly future, celebrating the marvels beneath our feet, and harnessing them for unparalleled home comfort

Enhanced Air Sealing and Insulation: The Shielding Mantle of Our Homes

When envisioning a perfect abode, we often focus on aesthetic elements: a charming façade, spacious interiors, or perhaps a manicured garden. Yet, one of the core pillars of a comfortable home often goes unnoticed - its building envelope. This uncelebrated hero, when optimized through enhanced air sealing and insulation, acts as the protective shield, holding the fort against external weather whims.

Unmasking The Building Envelope

At its core, the building envelope encompasses all the exterior components of a house - walls, windows, roof, foundation, and more. This physical separator between the conditioned and unconditioned environment determines how well your home can hold or repel heat.

Air Sealing: Think of air sealing as plugging the unseen chinks in your home's armor. Tiny gaps and cracks, which might appear insignificant, cumulatively allow a significant volume of air to seep in or out. By identifying and sealing these vulnerabilities, we prevent unintended air exchange with the outside.

Insulation: If air sealing is about plugging gaps, insulation is about adding layers of protective padding. Insulating materials act as barriers, slowing the rate of heat transfer. Whether it's the biting chill of winter or the scorching summer sun, insulation ensures that your home remains an oasis of comfort.

Basking in the Benefits

Consistency is Key: One of the first benefits homeowners notice post-enhancement is the consistent temperature. Rooms no longer have those pesky cold spots in winters or warmth-trapped corners in summers. The entire living space feels uniformly comfortable.

Reducing the HVAC Hustle: It's like a symphony where every element plays its part. When the building envelope does its job efficiently, the HVAC system doesn't have to overexert. It operates fewer cycles, maintains temperatures with ease, and, in essence, enjoys a longer, more efficient life. It's akin to having a car that you don't have to rev up constantly because it runs smoothly with minimal effort.

Slicing Through the Energy Bills: The math here is simple. Less work for the HVAC means reduced energy consumption. And when your heating or cooling system isn't working overtime, you'll see a marked reduction in monthly bills. It's a win for your wallet and the environment.

Enhancing Indoor Air Quality: An improved building envelope means fewer uncontrolled paths for air to enter or exit. This reduction limits the ingress of outdoor pollutants, allergens, and other contaminants. So, every breath you take indoors becomes a tad purer.

Humidity Control: Proper sealing and insulation also curtail moisture problems. By controlling the unbridled exchange of outdoor and indoor air, it becomes easier to manage indoor humidity levels, creating a healthier living environment and reducing risks of mold growth or structural damage.

Rolling Up Your Sleeves: A DIY Perspective

- **Start with an Audit**: Begin by conducting an energy audit. This helps in identifying the weak spots - areas where air leaks are prominent or insulation is lacking.

- **Sealant and Weatherstripping**: Armed with caulk, sealant, or weatherstripping, target the identified gaps around windows, doors, and other openings. Remember, it's the cumulative effect of sealing all these minor gaps that creates a significant difference.
- **Insulation Adventure**: For insulation, start with the attic, a common culprit for heat loss. Rolled batts or blown-in insulation, especially those made of fiberglass or cellulose, are popular choices. Walls and basements come next, and for the intrepid DIYer, there are numerous tutorials and guides available.
- **Safety First**: While this is an engaging endeavor, remember to take safety precautions. Wear gloves, eye protection, and a mask, especially when dealing with insulating materials.

But a word of caution: while minor sealing tasks can be tackled solo, for extensive work, especially insulation, consider getting professionals onboard. Their expertise ensures that the job's done right, and often, with materials and techniques that offer superior results.

In essence, enhanced air sealing and insulation are about reinforcing our homes' protective embrace. It's a step beyond mere walls and roofs; it's about creating a shield, a mantle that guards against the vagaries of weather, ensuring that the hearth remains warm, the rooms stay cool, and the inhabitants dwell in comfort.

The Role of Sustainable Refrigerants: Steering Clear of the Environmental Pitfalls

In the annals of HVAC history, a series of evolutions have constantly redefined our approach to cooling and heating. However, among these leaps, the transition towards sustainable refrigerants stands tall. Not just because it represents an engineering marvel, but because it echoes our collective consciousness to gravitate towards eco-friendly choices. Let's decode the shift and understand why it matters.

Chilling History: The CFC and HCFC Era

To appreciate the current momentum, a trip down memory lane is in order. Chlorofluorocarbons (CFCs) and Hydrochlorofluorocarbons (HCFCs) once dominated the refrigerant landscape. Why? Well, they were effective, stable, and seemingly ideal for a host of cooling applications. However, beneath the efficient façade lurked a darker side. CFCs and

HCFCs, when released into the atmosphere, began depleting the vital ozone layer. Picture this layer as Earth's aviator sunglasses, shielding us from the harsh ultraviolet rays of the sun. And with every leak or improper disposal of these refrigerants, those sunglasses developed a crack.

The Green Switch: Journey towards Sustainable Refrigerants

As the environmental impacts of CFCs and HCFCs came to light, the global community recognized the urgency to pivot. Enter the era of sustainable refrigerants. But what exactly makes a refrigerant "sustainable"?

- **Ozone-Friendly**: At its core, a sustainable refrigerant has little to no ozone-depletion potential. It's respectful of the ozone layer, ensuring our natural UV protection remains intact.
- **Low Global Warming Potential (GWP)**: Beyond the ozone, sustainable refrigerants also look at the bigger picture—minimizing their global warming potential. Simply put, they're designed to have a lower propensity to trap heat in our atmosphere, thus playing a part in combating climate change.

Diving into the Benefits

- **Protecting the Ozone Layer**: The most obvious perk? The ozone layer heaves a sigh of relief. With the phasing out of CFCs and HCFCs, and the embracing of sustainable refrigerants, we are taking concrete steps to halt, and hopefully reverse, the damage to this essential shield.
- **Combatting Climate Change**: High GWP refrigerants intensify the greenhouse effect, trapping heat within our atmosphere. Sustainable alternatives have a reduced GWP, ensuring that, in the event of a leak or release, their contribution to global warming is minimal.
- **Operational Efficiency**: It's not just about the environment. Many modern sustainable refrigerants offer superior thermodynamic properties. What's that in layman's terms? Your HVAC system might actually run more efficiently, consuming less energy while delivering equivalent, if not better, performance.

- **Safety Boost**: Several sustainable refrigerants have lower toxicity levels and are less flammable. This makes them safer in household settings, reducing risks associated with leaks or exposure.
- **Economic Incentives**: As the world marches towards green alternatives, many regions offer incentives, rebates, or tax breaks for systems using sustainable refrigerants. This can make eco-friendly choices also wallet-friendly in the long run.

While the topic may sound complex and technical, even the everyday homeowner can play a role in this transition.

- **Educate and Ask**: When scouting for a new HVAC system, or even during maintenance, ask about the refrigerant being used. Equip yourself with basic knowledge about sustainable options like R-32, R-410A, or R-134a, among others.
- **Maintenance is Key**: Even the most sustainable refrigerant can have an environmental impact if it leaks. Regular HVAC check-ups, ensuring seals and components are intact, can prevent unintended releases.
- **Recycle Right**: If replacing an older unit, ensure the refrigerant is properly recovered and recycled. Many service providers offer this, ensuring harmful refrigerants don't just get vented into the atmosphere.
- **Advocate**: Join local community groups or online forums to spread the word about sustainable refrigerants. As more people become aware and demand eco-friendly options, manufacturers and service providers will be nudged to align with these preferences.

When we think of environmental action, images of sprawling forests or vast oceans might come to mind. However, sometimes, significant changes come from seemingly inconspicuous sources – like the refrigerant in our air conditioner or refrigerator. The shift towards sustainable refrigerants isn't just a technical transition; it's a testament to our evolving ethos, recognizing that every choice, no matter how microscopic, ripples into a broader environmental impact. By choosing, using, and advocating for sustainable refrigerants, we're not just cooling our homes; we're nurturing a cooler, healthier planet.

Best Practices for Homeowners and Businesses: Optimizing HVAC Efficiency

An efficient HVAC system isn't just about advanced tech or the latest models; often, it's the everyday practices that keep the system running smoothly. If you're a homeowner or business, here's a distilled guide to elevating your HVAC game.

Regular Maintenance: A Stitch in Time...

- **Filter Vigilance**: A clean filter is an HVAC's best friend. Whether you're cooling down or heating up, ensure filters are replaced or cleaned every 1-3 months. This tiny act prevents your system from unnecessary labor, extending its lifespan.
- **Outside Unit Love**: Your outdoor unit, though sturdy, hates clutter. Periodically check for overgrown plants, debris, or miscellaneous items around it. A clear space ensures better airflow and efficiency.
- **Duct Duties**: Schedule an annual duct inspection. Clean ducts mean pure air and a system that doesn't strain to push air through.

System Upgrades: The Future is Efficient

- **High SEER Seeker**: If your system hails from a bygone era, think about upgrading. Newer models come with a superior Seasonal Energy Efficiency Ratio (SEER), translating to more bang for your energy buck.
- **Smart Thermostats, Smarter Choice**: They learn, adapt, and ensure you're comfortable without frittering away energy. A worthy investment, smart thermostats are like having a personal assistant for your HVAC.
- **Zone In on Zoning**: Got a sizeable property? Zoning might be your golden ticket. It allows different areas to have individual temperature settings. So, unused spaces won't guzzle energy.

Behavioral Practices: Mindfulness Matters

- **Tweak the Temp**: A minor adjustment in temperature, a degree up during summer or down in winter, can lead to substantial energy savings. It's a simple act with a pronounced impact.
- **Harness Your Window Power**: The sun is both a friend and foe. Use blinds to block out the scorching summer heat, but in winter, let those rays in to naturally warm your space.

- **Fans for the Win**: Ceiling fans can aid in even air distribution, making rooms feel comfortable at higher thermostat settings in summer. But remember, fans cool people, not rooms. Ensure they're off when the room is vacant.
- **For Businesses - Timing is Everything**: If feasible, businesses can align heavy HVAC use with off-peak hours. It could be kinder to your wallet, especially in areas with varied electricity pricing based on demand.

Armed with these practices, homeowners and businesses alike can cultivate an environment where comfort meets efficiency, and where the HVAC system operates at its peak potential.

CHAPTER 8

SAFETY FIRST!

Preventing Accidents and Problems: The Power of Proactivity

Ah, the joys of a smoothly running HVAC system - the unsung hero that keeps our living spaces comfortable. Yet, much like a car or any piece of machinery, a neglected HVAC system is an accident waiting to happen. So, let's dive right into why and how to ensure our HVAC systems remain the silent, efficient champions they were designed to be.

Regular Maintenance: A Non-Negotiable for Longevity and Safety

Imagine a scenario: It's the peak of winter, snow piling up outside. You're anticipating the warm embrace of your heater... and it fails. Frustrating, right? Now, consider a summer counterpart: scorching heat, but your air conditioner decides it's vacation time. Both instances aren't just about discomfort; they can be about potentially dangerous extremes.

Routine Checks to Prevent System Failures:

HVACs, under all their metal and wiring, are like us in a way; they benefit from regular health check-ups. Here's a blueprint for routine checks:

Monthly: Inspect filters. A clogged filter can hamper efficiency and force the system to work harder. Changing or cleaning them ensures unobstructed airflow and reduced strain on the system.

Quarterly: Look over the external components. The outside unit should be free of debris, plants, or any obstacles. Also, make sure drain pans and condensate drains are clear of blockages.

Annually: This is where professionals come in. An annual check by an HVAC specialist is like a thorough medical exam. They can check refrigerant levels, inspect electrical components, verify system controls, and ensure all moving parts are well-lubricated.

Importance of Addressing Minor Issues Before They Escalate:

Minor issues have an uncanny knack for snowballing into mammoth problems if left unchecked. Here's why addressing the little glitches is paramount:

Cost-Effectiveness: A stitch in time saves nine, they say, and with HVACs, this couldn't be truer. Addressing minor issues as they arise can save you from a hefty repair bill down the line.

System Longevity: Consistent minor tweaks and repairs can significantly extend your HVAC's lifespan. Think of it as providing regular nourishment to ensure it remains fit and running for the long haul.

Safety: This can't be emphasized enough. A malfunctioning HVAC isn't just a comfort issue. Faulty wiring can be a fire hazard, and a compromised system can risk carbon monoxide leaks, which are lethal.

Efficiency Maintenance: Addressing small issues regularly ensures that the system runs at optimal efficiency. This translates to consistent indoor temperatures and, as a bonus, reduced energy bills.

To draw an analogy, think of your HVAC system like teeth. Regular brushing (routine checks) keeps them healthy. But you still need that dentist visit (professional maintenance) to catch and address deeper issues. Ignoring a minor cavity (small issue) can lead to painful and costly root canals (major repairs).

Safe Installation Practices: Getting It Right the First Time!

"First impressions last a lifetime!" – We've all heard this saying, and it couldn't be more accurate when it comes to the HVAC world. Installation is the first impression your system gets of its new home. Now, do you want it settling in smoothly or experiencing the 'move-in' blues? Let's dive into making sure your HVAC system feels right at home from the get-go.

Employing Certified Professionals: The HVAC Whisperers

Picturing an HVAC installation might bring up images of a few wires here and a couple of screws there. But, in reality, it's so much more intricate. Think of it like a symphony, every part needs to be in tune, and the maestro that ensures this harmony is your certified professional. Here's why you absolutely, unequivocally need them:

Education & Training: These pros have had hours of training, learning the ins and outs of various HVAC models. Think of them as the HVAC whisperers—they just know what the system wants.

Equipped with the Right Tools: A certified technician comes armed, not just with knowledge, but with the precise tools needed for the task. This ensures that every component is fitted exactly as it should be, minimizing future issues.

Up-to-Date Knowledge: HVAC technologies are constantly evolving. Certified professionals stay updated with the latest trends and technologies, ensuring your system benefits from the latest best practices.

Ensuring Proper Installation: The Keystone of Longevity

Alright, you've got your certified professional, kudos! Now, let's delve into the nitty-gritty of ensuring a top-notch installation.

Correct System Sizing: One size doesn't fit all, especially in the HVAC world. A system that's too large will cycle on and off frequently, leading to energy wastage and uneven temperatures. Conversely, an undersized system will constantly struggle to maintain the desired temperature. Your professional will carry out calculations to determine the optimal system size for your space.

Accurate Refrigerant Levels: It's not just about filling it up and calling it a day. The refrigerant level in an HVAC system has to be just right. Too much or too little can reduce the system's efficiency and even lead to component damage over time.

Secure & Sealed Ductwork: The ducts are the highways for your conditioned air. Any leaks or poor connections mean your precious cold (or hot) air is escaping, making the system work harder. A professional ensures these ducts are sealed tight.

Electrical Connections Check: Loose wires or poor connections aren't just inefficient—they can be fire hazards. Your certified technician will ensure that all connections are tight, and components receive the correct amount of power.

Proper Drainage: The condensate drain needs special attention. A proper installation ensures that this drain is not only functioning but also directed to an appropriate place, preventing potential water damage or mold growth.

Thermostat Calibration: Your thermostat is the command center of the HVAC system. An improper calibration could mean the system thinks it's colder or hotter than it actually is. Proper installation ensures the thermostat reads accurately and communicates effectively with the system.

Awareness of Electrical Components: Navigating the Electrical Labyrinth Safely

Ah, the electrifying world of HVAC! Literally. Nestled within your HVAC system are a series of circuits, wires, and switches that make it hum, whistle, and sometimes, if not taken care of, sputter in protest. Now, before you get all "I got this" and dive hands first into this intricate web, let's understand the ground rules. Electrical safety isn't just a buzzword-it's the difference between a smooth, hazard-free HVAC operation and a potential mishap.

The Golden Rule: Power Off, Always!

Remember playing 'Freeze!' as a kid? In the world of HVAC electrical components, that game is a must, but with a slight tweak: instead of shouting 'Freeze!', you simply turn the power off. Here's why it's absolutely paramount:

Your Safety First: This isn't a scare tactic, but basic electricity 101. A live circuit can deliver a shock. Depending on the current and voltage, this can range from a mild jolt to something far more serious.

Protecting the System: It's not just about you; it's also about your HVAC. Tinkering with electrical components while they're powered can cause short circuits, blow fuses, or damage sensitive parts.

Clearer Inspection: With the power off, you can thoroughly inspect components without the hum or buzz of operation. This lets you spot any anomalies or issues that might be overlooked otherwise.

All Wired Up, But Make It Safe: The Insulation Game

It's a wire's world in there, and every wire has a story and a path. But more importantly, every wire has its coat-its insulation. Think of insulation as the winter jacket you don for that chilly morning walk; it's protective, snug, and ensures the cold (in this case, electricity) doesn't seep out. Let's untangle the reasons for proper insulation:

The Safety Buffer: Insulation acts as a barrier, ensuring the electric current remains within the wire. This not only protects you from potential shocks but also prevents accidental electrical fires that can be sparked by exposed wires.

Efficient Operation: Electrical leaks, although invisible to the naked eye, can strain the system. Proper insulation ensures all the power goes where it's supposed to, guaranteeing efficient system operation.

Protecting Against External Factors: In the snug world of an HVAC unit, wires can often come in contact with moving parts, water, or other elements. Insulation safeguards the wire from such external interferences, which could otherwise compromise its functionality.

Longevity of Components: Properly insulated wires reduce the risk of short circuits, electrical overloads, or ground faults. This in turn enhances the lifespan of various electrical components in the HVAC system.

While it might be all too tempting to channel your inner electrician superhero, it's essential to remember that knowledge is power (pun intended). Knowing when to shut off the power and ensuring that every wire is snug in its insulation jacket is crucial. While these steps might seem minor in the grand scope of HVAC maintenance, they're foundational in ensuring safety and efficiency.

Proper Ventilation: Ensuring Safe Air Circulation

Ventilation plays a fundamental role in the operation and efficiency of HVAC systems. Ensuring that exhaust gases are correctly vented outdoors and that the ventilation system remains free from obstructions is paramount to the safety and efficiency of the entire setup.

Directing Exhaust Gases Outdoors

When HVAC systems operate, especially those dependent on combustion processes, they produce exhaust gases. These gases need to be directed outside for several reasons:

Health and Safety: Combustion processes can produce harmful gases such as carbon monoxide, which is dangerous when accumulated indoors. Directing these gases outdoors safeguards the health of the building's occupants.

Operational Efficiency: An HVAC system with a clear path for its exhaust can operate more efficiently. Proper venting ensures there's no buildup of heat or gases that could strain the system or lead to malfunctions.

Component Longevity: Ensuring that gases are vented appropriately can help prevent the accumulation of corrosive substances inside the system, which could potentially damage components and reduce their lifespan.

Importance of Regular Checks in Ventilation Systems

To ensure that ventilation systems operate correctly and that exhaust gases are appropriately directed outside, regular inspections are necessary:

Identifying Blockages: Over time, vents can become obstructed by debris, dust, or even nests built by small animals. Regularly inspecting these pathways ensures that any blockages are identified and cleared in a timely manner.

Checking for Leaks: External factors, age, or general wear and tear can lead to leaks in the ventilation system. Regular checks help in spotting these leaks early on, ensuring that exhaust gases are directed appropriately and preventing any energy wastage.

Maintaining Indoor Air Quality: A well-maintained ventilation system can help in reducing the buildup of pollutants, mold, or dust inside the vents, contributing to better indoor air quality.

Avoiding Backdrafting Issues: Proper ventilation checks can help identify issues where exhaust gases might be pulled back into the building instead of being vented out, a phenomenon known as backdrafting. This is crucial, as backdrafting can introduce harmful gases back into the indoor environment.

Ensuring a Safe System for the Whole Family:

Childproofing: Safeguarding Your HVAC System from Little Explorers

As any parent or guardian knows, children are naturally curious. This curiosity often leads them to explore every nook and cranny of a home. Given that HVAC systems can present potential dangers when tampered with, it's imperative to take precautions to keep these young explorers safe.

Childproofing Overview

Childproofing involves more than just ensuring that your little ones don't have access to electrical outlets or sharp objects. When it comes to your HVAC system, the goal is twofold: protect the child from the system and the system from the child. Kids might find the whirring sounds and the warm or cool breezes emanating from the vents fascinating, but there are components and settings that, if tampered with, can jeopardize both their safety and the functionality of the system.

Protective Barriers: A First Line of Defense

One of the simplest yet most effective ways to childproof your HVAC system is by installing barriers:

Around Outdoor Units: The outdoor units can be intriguing to children because of their size and the noises they make. Placing a protective fence or barrier around these units can prevent kids from poking fingers into moving parts or dropping objects inside.

Indoor Vents and Radiators: Consider using vent and radiator covers that prevent small hands from accessing the interior parts but still allow for efficient air circulation.

Lockable Covers: Safety and Settings Intact

Children love to turn dials, press buttons, and flip switches. While it might be fun for them, it can play havoc with your HVAC system settings:

Thermostat Covers: Transparent, lockable covers can be installed over thermostats. This allows adults to see and adjust settings while ensuring that those settings aren't changed accidentally during a child's play.

Lockable Access Panels: Some HVAC components, like central air systems or larger radiators, might have access panels that children could potentially open. Ensure these panels have locks or childproof latches to prevent unintended access.

Remote Controls: Out of Sight, Out of Mind

Today, many modern HVAC systems come with remote controls for ease of operation. These remotes are often just the right size for children to play with:

Store in Elevated Places: Make it a habit to keep remotes on higher shelves or in drawers out of a child's reach.

Teach & Guide: As children grow, teaching them the importance and the 'do's and 'don'ts' concerning the remote can be beneficial. This not only safeguards the HVAC system but also instills a sense of responsibility in them.

While childproofing might seem like a continuous task, especially as children grow and find new ways to explore, it's a worthwhile endeavor. Protecting both children and valuable household systems is of paramount importance. By installing barriers, using lockable covers, and ensuring that remote controls are stored properly, parents and guardians can ensure a safer environment where kids can play and explore without jeopardizing their safety or the efficiency of the HVAC system.

Pet Safety: Guarding Your Furry Friends Against HVAC Hazards

Our homes are filled with members who walk on four legs and offer unconditional love. Pets, be it a frisky feline or a loyal canine, bring joy to our lives. As much as we love them, it's essential to realize that HVAC systems, often ignored by the human members of the family, can pose potential threats to these furry friends. Ensuring their safety is just as crucial as maintaining the efficiency of your heating and cooling systems.

Pet Safety Overview

Dogs, cats, and other household pets might not share a child's innate curiosity for poking and prodding, but they have their unique set of behaviors that can interact with HVAC systems. Whether it's a cat that loves to nap on a warm vent or a dog that sees an outdoor unit as a potential threat to be barked at, there are risks and challenges that pet owners need to address.

Fencing Off Outdoor Units: Creating Safe Boundaries

Outdoor units, with their humming sounds and fan blades, can be particularly interesting, or even threatening, to pets:

Physical Barriers: A simple fence or enclosure around your outdoor HVAC unit can serve multiple purposes. It can prevent pets from urinating on the unit, a behavior that can corrode and damage its parts. Moreover, it safeguards curious pets from getting too close to moving parts and electrical components.

Safe Distance: Ensure that the barrier you place around the unit leaves enough space (typically 2-3 feet) for adequate airflow and for technicians to service the unit when required.

Checking for Pet Hair Blockages: A Hairy Situation

The shedding of fur is natural for many pets, but when this hair finds its way into vents and filters, it can pose a challenge:

Regular Cleaning: Regularly sweep and vacuum areas near vents and intakes. This will reduce the amount of hair that gets sucked into the system.

Filter Checks: Pet owners should consider checking and replacing their HVAC filters more frequently than others. A clogged filter, heavy with pet hair, makes your system work harder and can diminish indoor air quality.

Securing Floor Registers: No Paws Allowed

Floor registers can become inadvertent traps for unsuspecting paws or curious noses:

Grate Selection: Opt for registers with smaller gaps in their grates. This prevents small paws or claws from getting caught.

Secure Placement: Ensure that the registers are securely fastened. Pets might inadvertently dislodge a loose register, exposing them to the ductwork beneath.

Monitor Behavior: Some pets might view vents as excellent places to stash toys or hide treats. Monitor your pets and train them to avoid these behaviors, ensuring that your ducts remain clear and free-flowing.

In the vast tapestry of pet ownership, HVAC systems might seem like a minor thread. However, as with all household systems, a little foresight, regular maintenance, and awareness can go a long way in ensuring that our beloved pets stay safe and our homes remain comfortable. The steps mentioned might appear simple, but they can make a significant difference in the life and safety of our furry companions.

Air Quality Maintenance: Breathing Easier in Your Space

In the grand landscape of home management, air quality might not always be the first priority. Yet, the quality of the air we breathe indoors can significantly impact our health, comfort, and overall well-being. As homeowners, ensuring that our indoor environment remains free of harmful pollutants is paramount. Dive in to understand the significance of maintaining stellar air quality and the measures one can adopt to achieve this goal.

The Vitality of Clean, Breathable Air

Ever paused to think about how much time you spend indoors? For many, this figure easily crosses 80% of their day. Indoor air, often believed to be cleaner, can be 2-5 times more polluted than outdoor air. This makes maintaining its quality not just a luxury, but a necessity.

Pollutants, allergens, and pathogens that thrive in poorly maintained environments can trigger asthma attacks, allergies, and other respiratory issues. It's not just about preventing sneezes or coughs; it's about safeguarding the health of everyone living under that roof.

Stepping Up the Game with Regular Filter Changes

The first line of defense in any HVAC system is the humble air filter. It works tirelessly, trapping dust, pollen, and a myriad of airborne particles:

Scheduled Swaps: It's recommended to change out HVAC filters every 30-90 days, depending on usage and whether there are pets in the home.

Upgrading Your Arsenal: Standard filters do a decent job, but for homes in areas with higher pollution or for families with allergy sufferers, consider upgrading to HEPA (High-Efficiency Particulate Air) filters. They can trap particles as small as 0.3 microns, ensuring cleaner air circulation.

Air Purifiers: Your Home's Silent Guardians

For those especially concerned about indoor air quality or living in densely populated urban areas, air purifiers can be a game-changer:

How They Work: Air purifiers draw in room air, pass it through a series of filters, and release clean air. Over time, they can significantly reduce the number of pollutants indoors.

Choosing the Right One: Not all purifiers are created equal. Look for ones that cater to your specific needs - be it pet dander, pollen, or smoke.

Combatting the Unseen: Mold and Pollutants

Moist, dark spaces can often become breeding grounds for mold. Similarly, household activities like cooking can introduce pollutants:

Routine Inspections: At least once a year, especially before humid seasons, inspect your home for damp spots or any signs of mold growth.

Ventilate: Proper ventilation, especially in areas like the kitchen and bathroom, can significantly reduce the buildup of pollutants. Ensure exhaust fans work efficiently and consider cracking a window occasionally to let fresh air circulate.

Houseplants to the Rescue: Some houseplants, like the Spider plant or Peace Lily, can act as natural air purifiers, removing certain toxins from the air. It's a green, aesthetically pleasing way to enhance air quality.

Keeping the air in our homes clean is akin to providing a protective shield against numerous health issues. With these strategies in hand, you're well on your way to creating a space that's not just comfortable but also safe and healthy.

Emergency Protocols: Safeguarding Your Home and Loved Ones

When it comes to our homes, it's easy to operate under the assumption that all is well until something suddenly goes awry. HVAC systems, while designed for durability, aren't exempt from the occasional hiccup or major malfunction. Instead of being caught off guard, let's ensure that you're prepared. In the face of HVAC emergencies, a swift, knowledgeable response can be a lifesaver - literally.

Understanding the Need for Emergency Protocols

You've invested time, money, and effort in getting the best HVAC system. Yet, like any machine, it can falter. From a simple glitch causing temperature discomfort to serious malfunctions potentially leading to fires or gas leaks, the spectrum of what might go wrong is vast. Having a predetermined plan is not about expecting the worst but ensuring safety amidst unforeseen challenges.

1. Mastering the Art of System Shutdown

Before diving deep into problem-solving, your priority in any emergency should be safety. And that often means knowing how to shut off your HVAC system:

Locate the Switch: Every HVAC unit comes with a shut-off switch. Familiarize yourself with its location, usually found on a wall near the unit or inside the unit's access panel.

Power Down: In case of any suspicious activity-like strange noises, burning smells, or unexpected shutdowns-turn the system off immediately. This can prevent further damage and offer a buffer period to evaluate the issue.

2. Keeping Essential Contacts Within Arm's Reach

In the heat of the moment-pun unintended-you don't want to be rummaging through directories or scouring the internet:

Primary HVAC Service Provider: This should be your first call. They'll have details on your system's specifics, warranty, and service history.

Emergency Services: For severe emergencies, especially if you suspect a gas leak, dial emergency services without hesitation.

Dedicated Space: Consider maintaining a dedicated space in your home - perhaps on the refrigerator or pinned to a notice board - where these emergency numbers are visibly displayed.

3. Engaging in Family Safety Drills

Emergencies can be disorienting, especially when they catch you off guard. Regular safety drills can help:

Map Escape Routes: Every family member should know the quickest, safest way out of the house from different points. This is crucial, especially if there's a risk of fire.

Safe Zones: Designate a safe gathering point outside the home. It ensures that everyone can be accounted for and prevents anyone from re-entering a potentially hazardous area.

Role Allocation: Assign specific roles. While one person shuts down the HVAC system, another might be responsible for calling the technician, and yet another ensures pets are safe.

Drill Regularity: Conduct these drills every few months. Make it a household routine, ensuring newcomers, guests, or young children are equally informed.

Having an HVAC emergency protocol isn't about dwelling on the negatives but about empowerment. It's about equipping yourself and your loved ones with the knowledge and tools to handle challenges with calmness and efficiency. So, take the time today to familiarize, plan, and practice. After all, in the face of emergencies, preparation isn't just power - it's safety.

Educating Family Members: Building a Safe HVAC Haven

In the grand tapestry of household activities, HVAC systems might seem like just background machinery - humming quietly and doing their job. But just as we educate our family members about fire safety or the correct way to operate kitchen appliances, it's crucial that everyone in the home understands basic HVAC safety protocols.

Why HVAC Safety Education Matters

At its core, the HVAC system is a powerful piece of equipment. Its purpose is to regulate your home's climate, making it a vital component especially during extreme weather conditions. But with great power comes great responsibility - mishandling or misunderstandings can lead to unnecessary hazards.

Prevention Over Cure: We've all heard it, and it applies here perfectly. Educating family members about potential risks and correct handling can prevent avoidable emergencies.

Shared Living, Shared Responsibility: Everyone uses the HVAC system. Whether it's adjusting the thermostat, changing filters, or simply ensuring vents aren't blocked, it's a collective responsibility.

Empowerment Through Knowledge: When everyone in the household knows what to do, especially during unexpected situations, it reduces panic and promotes effective action.

Hosting Effective Family Safety Meetings

Now that we understand the significance of HVAC safety, let's delve into the how-to. Hosting regular family meetings can be an excellent way to ensure everyone's on the same page.

Set a Schedule: Just as you would with a family dinner or movie night, set a date. This isn't a one-off task. Regular refreshers, especially as seasons change, can be helpful.

Demystify the Machinery: Start with a basic overview of the HVAC system. Younger family members might find it fascinating! Show them the primary components like the thermostat, vents, and main unit. Explain the system's function in simple terms.

Outline Do's and Don'ts: Clearly discuss the safety protocols. This can include guidelines like:

- Not obstructing vents.
- Maintaining a certain distance from outdoor units.
- The right way to adjust settings on the thermostat.

Open the Floor for Questions: This is crucial. Encourage family members to ask questions, no matter how basic they seem. Addressing these queries not only clarifies doubts but reinforces the importance of safety.

Role Play: Engage in mini-drills or role-playing scenarios, especially with younger members. It can make the learning process interactive and memorable.

Resources at Hand: Consider creating a small, easy-to-understand manual or guide that's available for reference. Place it near the HVAC unit or thermostat.

Feedback is Golden: After the meeting, take feedback. Find out if there are areas that need more clarity or if there are specific concerns that weren't addressed.

The beauty of making HVAC safety a family affair is that it creates a shared sense of responsibility. The system no longer remains a complex machine that only one person in the house understands. Instead, it becomes a communal asset, guarded and maintained by all. After all, when it comes to safety, there's no better approach than collective vigilance and shared knowledge.

CHAPTER 9
THE ECONOMIC ASPECT OF HVAC

Initial Investment and Long-term Returns: Evaluating Costs and Benefits

Choosing the Right System: The Nexus of Upfront Expenditure and Future Savings

HVAC systems, forming the crux of a building's climate control, present a significant decision point for homeowners and businesses. Understanding the delicate interplay between initial costs and potential long-term savings can guide the decision-making process.

Balancing Immediate Outlays with Anticipated Savings

Initial Price vs. Long-term Costs: Lower upfront costs can be tempting. However, initial savings on a less efficient HVAC system may be negated by higher operational expenses. These operational costs manifest in the form of increased energy consumption, frequent maintenance requirements, and potentially earlier replacement needs.

Efficiency Equates to Monetary Savings: A high-efficiency HVAC system, while possibly commanding a higher purchase price, operates using less energy. The reduced energy usage translates directly to monthly savings on utility bills. For instance, if an efficient system conserves even 15% on monthly energy bills, the aggregated savings over several years can offset the initial price difference.

Financial Incentives: Many local governments and utility companies reward the installation of energy-efficient systems through tax credits, rebates, or reduced tariffs. Such incentives can help mitigate the higher initial costs of efficient HVAC systems.

The Indispensable Role of Efficiency

Environmental Stewardship: High-efficiency HVAC systems consume less energy, which has a direct positive impact on the environment. Reduced energy consumption means fewer carbon emissions from power plants, contributing to a lower carbon footprint.

Uniform Performance: One hallmark of an efficient HVAC system is its ability to maintain consistent temperatures across different areas. Such uniformity ensures that the system doesn't overwork to rectify temperature imbalances, leading to a decreased likelihood of system failures or malfunctions.

Durability: As a rule of thumb, efficient HVAC systems are built for longevity. They typically incorporate robust components that withstand the rigors of daily operation. Fewer operational strains translate to a longer functional life, pushing replacement needs further into the future.

Property Valuation: Efficient HVAC systems can enhance the overall value of a property. When selling a residential or commercial space, potential buyers often factor in the efficiency of installed systems. A top-tier HVAC system signals reduced future expenses, making the property more appealing to discerning buyers.

In the realm of HVAC systems, decision-makers must adopt a forward-looking approach. The focus should pivot from mere initial costs to a broader perspective that considers long-term operational expenses and benefits. A robust, efficient HVAC system, while perhaps demanding a higher initial outlay, can offer tangible savings and benefits in the ensuing years. Making a well-informed decision can ensure optimal climate control without undue financial strain.

Installation Costs: Understanding the Essentials

Deciphering Installation Costs

Embarking on an HVAC journey requires homeowners and businesses to account for multiple costs, with installation being a significant component. Knowing what factors play into this expense can provide clarity on where your money goes and ensure you make an informed decision.

Factors Influencing Installation Costs

Size and Capacity of the System: A basic principle is at play here; a larger system designed to serve a more extensive space often incurs a higher installation cost. This relates directly to the amount of labor and materials needed for the setup. But, it's critical to remember that a system's size must be tailored to the space it's meant to serve. Too big or too small, and you could face efficiency issues.

Complexity of Installation: Every building or home brings its own set of challenges. Some might already be equipped with functional ductwork, while others could need a complete rework. Naturally, installations that are more complex in nature require more time and expertise, leading to a steeper cost.

Brand and Model Specifications: Some HVAC brands or specific models, especially those at the forefront of technology, might be pricier to install. They may offer advanced features that require a specialized touch during the setup.

Geographic Considerations: The region you reside in can also influence installation costs. Labor rates fluctuate across areas, with metropolitan locales typically charging more than their rural counterparts.

Existing Infrastructure Adjustments: If your property necessitates modifications like electrical upgrades to support the new system or needs removal of an old unit, these tasks add to the total installation cost.

The Imperative of Professional Installation

With various DIY tutorials available, one might ponder, "Why not attempt a self-installation?" While the sentiment is understandable, the stakes are high, and here's why professional installation is non-negotiable:

Accuracy Matters: HVAC setups are intricate by nature. Even a slight misstep during installation can set off a chain of operational issues. Professionals come prepared with the necessary training to ensure every piece is meticulously fitted and configured.

Guaranteeing Efficiency: An improperly installed HVAC system, no matter how advanced, won't deliver its promised efficiency. Entrusting professionals ensures your system operates at its intended capacity, translating to energy savings and optimal indoor comfort.

Safety Concerns: Installing an HVAC system brings its own set of hazards, from potential gas leaks to electrical complications. Trained technicians are adept at navigating these challenges, ensuring the end setup is secure and reliable.

Warranty Specifications: A critical aspect many overlook is the warranty clause. Several manufacturers mandate professional installation for the warranty to remain valid. Bypassing this requirement could leave you unprotected in the face of future system issues.

Extending System Life: A well-installed system naturally experiences fewer operational hitches. This reduces wear and tear, possibly extending the system's operational life and ensuring you get the most out of your investment.

In the realm of HVAC, cutting corners during installation isn't an option. The initial outlay for a professional setup paves the way for a system that's efficient, safe, and primed to serve you effectively over its lifespan.

Maintenance and Repair Budgeting: Understanding Your HVAC Costs

Navigating the Landscape of HVAC Care

When you've made the significant investment in an HVAC system, it's essential to comprehend the accompanying costs of upkeep. An HVAC system, while robust, requires attention to ensure its longevity and efficiency. Let's delve into the financial aspects of this responsibility.

Breaking Down Annual Maintenance Costs

Routine Checks: Industry professionals universally recommend two primary inspections annually: one preceding the heating season and another before the cooling season kicks in. These evaluations can vary in price, typically falling between $150 and $300 per annum.

Filter Replacements: The role of filters is pivotal. They filter out dust, pollen, and other potential pollutants, ensuring clean airflow. Over a period, they become less efficient due to accumulated debris, necessitating replacements. The annual expense for this component might oscillate between $20 and $120, contingent on the system and filter type.

Miscellaneous Repairs: HVAC systems, given their complexity, might require unforeseen repairs. Allocating a budget ranging from $150 to $500 yearly is a prudent strategy. The exact amount is influenced by factors such as the system's age and usage frequency.

The Economics of Preventive Maintenance versus Reactive Repairs

Preventive Maintenance:

Pros:

Cost Efficiency: By identifying potential issues early, preventive measures can negate the need for substantial repairs, ensuring cost-effectiveness in the long run.

Extended System Life: Regular upkeep ensures optimal system performance, which can elongate the equipment's lifespan.

Enhanced Efficiency: Focused maintenance of elements like filters, coils, and belts means the system operates at peak efficiency. This can translate to lower energy consumption and, by extension, reduced utility bills.

Cons:

Recurrent Costs: Some homeowners might view the periodic expenses associated with preventive maintenance as a financial burden.

Reactive Repairs:

Pros:

On-demand Expenditure: A section of homeowners might lean towards addressing issues only as they arise, thereby sidestepping periodic maintenance costs.

Cons:

Cumulative Expenditure: Ignoring small problems can lead to them mushrooming into larger issues, resulting in heftier repair bills.

Shortened System Life: A system subjected to continuous use without timely maintenance checks may have a reduced operational life.

Operational Interruptions: Adopting a wait-and-watch approach until a system fault arises can lead to inconvenient downtimes, affecting the comfort levels of a household.

A comprehensive view indicates that preventive maintenance is the more financially and operationally sound choice. By incurring consistent, planned expenses, homeowners can ensure system efficiency, elongate the equipment's lifespan, and potentially curtail substantial repair costs down the line.

Monthly Utility Costs and Energy Efficiency: Navigating Your HVAC Expenditures

Navigating through monthly utility bills might feel like a walk in a maze. Within these bills, the HVAC system plays a significant role. It's essential to understand how heating and cooling can directly influence these costs. Let's embark on this exploration.

Understanding Your Bill: Breaking Down the HVAC Component

Deciphering the HVAC Influence on Your Utility Bill

The utility bill you receive each month aggregates energy consumption from various sources. Of these, the HVAC system, responsible for maintaining the comfort of your premises, usually takes up a significant portion. Statistically speaking, HVAC activities can account for around 48% of the energy use in a standard U.S. home.

HVAC Energy Consumption in Detail

Kilowatt-Hours (kWh) Consumption: Your bill primarily revolves around the kWh, a standardized unit measuring energy over time. The energy consumption of your HVAC system, which varies based on its efficiency, operational hours, and climatic conditions, directly contributes to the kWh count.

Tiered Rates: Some utility providers employ a tiered rate system. Here, energy consumption beyond specific thresholds incurs progressively higher rates. An inefficient or overactive HVAC system can push your consumption into these higher tiers.

Time-of-Use Rates: In some regions, energy costs differ according to the time of day. During periods of peak demand, typically the warmer parts of the day, rates can be higher. If your HVAC is in high gear during these periods, it can amplify your costs.

Unraveling Terms & Charges

Utility bills may appear complex due to specific terminologies. Let's break these down for clarity:

Usage Charge: This segment reflects the total cost of the kWh your property consumed during the billing cycle. Especially during temperature extremes, a substantial chunk of this can be attributed to the HVAC system.

Demand Charge: While less common in residential billing, some utilities charge based on the maximum instantaneous electricity demand. This can be more pertinent to businesses or larger properties.

Fixed Charge or Service Charge: A constant fee covering the costs associated with infrastructure maintenance, metering, and other administrative processes. This charge remains the same, irrespective of HVAC energy consumption.

Taxes and Surcharges: Government-imposed fees that aren't a direct result of HVAC operations but do affect the overall bill.

Rate Schedule or Tariff Code: This code signifies your billing plan. Terms like "TOU" (Time-of-Use) or "Tiered" denote your utility's specific pricing structure. As previously explained, these structures can heavily influence HVAC-related costs.

Therms or CCF: If a gas heating system is in place, the bill might feature charges based on "therms" or "CCF." This represents the volume of natural gas consumed, and a spike in this figure can signify increased furnace usage.

Understanding the nuances of utility billing, especially the influence of HVAC operations, is paramount for homeowners and businesses. With this clarity, you can optimize HVAC usage, potentially leading to a balanced utility bill. Moreover, being aware of billing components helps in more efficient energy management, aligning with sustainable practices and economical expenditures.

Energy-Efficient Systems and Their Impact: A Deep Dive into Savings and Returns

In the bustling hubbub of our daily lives, many homeowners and business operators are always looking for effective ways to economize without sacrificing comfort. Enter energy-efficient HVAC systems—a powerful solution that not only helps the environment but also saves a pretty penny. Let's dissect how these systems make a monumental difference in monthly bills and the bigger picture of return on investments (ROI).

How Energy-Efficient HVAC Systems Drive Down Monthly Costs

1. Reducing Energy Consumption: At its core, an energy-efficient HVAC system is designed to utilize less energy while delivering the same, if not better, performance. This optimized operation translates directly to fewer kilowatt-hours (kWh) on your utility bill. With the HVAC system accounting for a significant portion of a household's energy use, even a small percentage reduction can lead to tangible savings.

2. Minimizing Wear and Tear: Energy-efficient systems often incorporate cutting-edge technologies that reduce the operational strain. This means components are less likely to wear out or malfunction, leading to lower maintenance costs.

3. Peak Performance during Peak Hours: Many regions employ Time-of-Use (TOU) billing, where energy rates surge during high-demand periods. Energy-efficient HVAC systems often integrate "smart" features, adjusting operations to off-peak hours, or running optimally during high-demand times, avoiding excessive costs.

ROI Calculations: The Road to Recouping Your Investment

While the initial investment in an energy-efficient HVAC system might be higher, it's essential to see this as planting a seed—an investment for future financial growth. Calculating ROI helps in quantifying this benefit. Here's how you can get a clearer picture:

1. Establishing a Baseline: Begin by determining your current monthly and annual HVAC-related costs with your existing system. This becomes your reference point.

2. Projected Monthly Savings: With the estimated efficiency percentages provided by manufacturers or installers, calculate the potential monthly savings. For instance, if your older system costs $200 a month to operate and the new system promises 20% energy savings, you can expect to save about $40 monthly.

3. Annual Savings: Multiply your monthly savings by 12 for a ballpark annual savings figure. Using the above example, this amounts to a substantial $480 yearly.

4. Total Cost of the New System: Factor in the purchase price, installation charges, potential rebates or incentives, and any financing costs to arrive at the total investment in the new HVAC system.

5. Crunching the ROI Numbers: Here's the fun part! To determine how long before your energy-efficient system pays for itself, divide the total system cost by your annual savings. If the new system costs $6,000, and you save $480 yearly, it would take roughly 12.5 years to fully recoup the investment via savings.

6. The Bigger Picture: While the above calculations offer a direct financial perspective, it's worth noting the intangible benefits. An energy-efficient system enhances property value,

offers more consistent indoor comfort, and contributes to a smaller carbon footprint. While these aren't directly calculable in ROI, they offer intrinsic long-term value.

Embracing energy-efficient HVAC systems is akin to steering a ship towards calmer waters. While the initial costs might feel like a hefty wave, the ensuing smooth sail of reduced monthly bills and the peace of knowing you're doing right by the planet makes it all worthwhile. Armed with a keen understanding of potential savings and ROI, homeowners and businesses can make informed decisions, ensuring a comfortable, cost-effective, and sustainable environment.

The Role of Insulation and Sealing in HVAC Efficiency

Insulation and sealing, although often overlooked, play pivotal roles in determining the efficiency and functionality of an HVAC system. Their primary purpose? To regulate the indoor environment, ensuring that the HVAC system doesn't have to overexert itself. This translates directly into energy savings and cost reductions.

The Mechanics of Insulation and Sealing

Insulation functions to limit the transfer of heat. In the colder months, insulation keeps the warmth inside, preventing it from escaping outwards. Conversely, during the hotter periods, insulation acts as a barrier against the external heat, stopping it from infiltrating indoor spaces. An HVAC system, when operating within a well-insulated environment, doesn't have to strain to maintain consistent temperatures, resulting in reduced energy usage.

Sealing, on the other hand, deals specifically with potential gaps and openings in the building's structure, such as those around doors and windows. Proper sealing ensures that the conditioned air, be it warm or cool, remains inside. In the absence of effective sealing, HVAC systems might continually work to compensate for the lost air, leading to increased energy consumption.

Enhancing Insulation and Sealing: Practical Tips for Homeowners

1. Energy Audits: One of the most straightforward steps homeowners can take is to conduct an energy audit. A professional assessment will identify areas where energy loss occurs, guiding you towards necessary insulation and sealing enhancements.

2. Choosing Insulation: There are multiple insulation types, including fiberglass, foam, and cellulose. The choice depends on the specific needs of the space. For instance, while fiberglass might be apt for walls, foam can be suited for areas around HVAC ducts.

3. Prioritize Windows and Doors: Begin by addressing the usual suspects. Employ weatherstripping for doors and windows. For visible gaps in walls or between floorboards, caulk can be applied for a tight seal.

4. Attic Attention: Heat rises, making attics prime areas of concern, especially during winters. Enhancing attic insulation can lead to noticeable energy savings.

5. Addressing Basements: Basements and crawlspaces, if neglected, can contribute to energy inefficiencies. Wall insulation in basements and vapor barriers in crawlspaces can mitigate these concerns.

6. Invest in Energy-Efficient Windows: Double-pane windows significantly diminish heat transfer. If replacing windows isn't feasible, homeowners can also explore the option of window films which reflect heat.

7. Seal Outlets and Switches: Outlets and switches, though small, can be sources of drafts. Simple foam gaskets, available widely, can be placed behind them to eliminate these minor air leaks.

8. Seal HVAC Ducts: Properly sealed ducts are essential for the efficient transfer of air. Using recommended materials like mastic sealant or metal tape ensures that the ducts remain leak-free.

A building, whether residential or commercial, functions much like an integrated system. Every element, from the walls and windows to the HVAC unit, plays its part. In this system, insulation and sealing are the unsung heroes. By optimizing these two elements, homeowners can achieve a balance of comfort and cost-effectiveness.

Smart Systems and Thermostats: Harnessing Advanced Controls for Economic Advantages

Modern living spaces, complemented by advancements in technology, are increasingly gravitating towards efficient systems that do more than just perform a singular function. One such marvel in the HVAC domain is the advent of smart systems and thermostats. These tools, driven by sophisticated algorithms and sensing technologies, offer homeowners an unprecedented level of control and efficiency. In this chapter, we'll deep dive into the economic benefits they bring to the table and how you can gauge potential savings from these optimized HVAC operations.

The Renaissance of HVAC Control: Smart Thermostats

A shift from the traditional, manual dials to thermostats that you can program or even control remotely using your smartphone signals a major leap in HVAC technology. But what truly makes them shine is their ability to learn and adapt to user behaviors and preferences.

Adaptive Learning: Many modern thermostats track user behaviors over time. For instance, if you consistently lower the temperature at 10 PM, the system will begin to do it automatically.

Remote Control: Caught in a sudden weather change while you're away? Smart thermostats allow users to adjust settings from anywhere, ensuring the home is at the desired temperature upon return, without wasting energy during the absence.

Energy Consumption Reports: These thermostats provide detailed reports on energy usage, offering insights into peak consumption periods, allowing homeowners to adjust accordingly.

Economic Advantages: The Monetary Wisdom Behind the Tech

At a glance, a smart thermostat might appear as just another gadget. However, its economic benefits are tangible and noteworthy.

Reduced Energy Wastage: Traditional thermostats lack the precision of their smart counterparts. With a smart system, the heating or cooling is fine-tuned to your needs, avoiding overexertion and thus reducing energy waste.

Peak Usage Awareness: With insights into peak energy consumption periods, homeowners can capitalize on off-peak rates (if available) by adjusting their usage patterns, potentially leading to lower utility bills.

Less Wear and Tear: An efficiently managed HVAC system experiences less strain, which can translate to fewer repairs and an extended lifespan for the equipment.

Estimating Savings: What Could You Save?

Pinning down exact numbers for savings can be tricky due to variances in utility rates, home sizes, and individual usage patterns. However, a general approach can provide a ballpark figure.

Baseline Data: Begin by analyzing your utility bills prior to the installation of a smart thermostat. This provides a clear picture of your consumption patterns.

Post-Installation Analysis: After using the smart thermostat for a few months, compare the new utility bills with the old ones. The difference often showcases the direct impact of smarter HVAC management.

Annualize the Savings: While monthly savings are encouraging, it's the annual figure that truly illustrates the economic advantage. Multiply your average monthly savings by 12.

Factor in External Elements: Remember to account for any significant external changes, such as unusually harsh weather conditions or changes in utility rates, as they can skew the savings data.

Fine-Tuning For Maximum Efficiency

While the installation of a smart thermostat can lead to passive savings, active engagement can push the envelope even further. By regularly reviewing energy consumption reports, adjusting settings for maximum efficiency, and staying updated on newer software versions or updates for your device, homeowners can truly harness the power of their smart systems.

In the grand narrative of HVAC advancements, smart thermostats and systems mark a chapter where technology melds seamlessly with user comfort and economic prudence. The immediate savings they offer, coupled with long-term benefits of efficient energy management, make them more than just a luxury—they're a wise investment for the modern homeowner.

Tax Credits and Rebates: The Economic Incentives for Energy Efficiency

The environmental benefits of energy-efficient systems are widely touted, but there's also a silver lining that homeowners frequently overlook: the economic incentives. As governments and organizations globally recognize the impact of energy consumption on the environment, they've instituted tax credits, rebates, and incentives to encourage homeowners and businesses to make greener choices. Let's delve into the specifics of these advantages and guide you on how to capitalize on these financial opportunities.

A Primer on Tax Credits, Rebates, and Incentives

Before diving into the 'how-to', let's first differentiate between these terms:

Tax Credits: A tax credit reduces the total amount of income tax you owe to the federal and/or state government. For instance, if you owe $3,000 in taxes but qualify for a $1,000 energy efficiency tax credit, your net payable tax drops to $2,000.

Rebates: Often provided by utility companies or manufacturers, rebates give you a refund on the cost of the equipment or service. If you buy an energy-efficient HVAC system for $5,000 and there's a $500 rebate, you'll receive that amount back, making your net expenditure $4,500.

Incentives: These are additional perks or bonuses, often from local governments or utilities, for making energy-efficient choices. They might include discounted services, extended warranties, or even free additional equipment.

Seeking Out Available Opportunities

Different regions and utilities offer varying incentives, so it's crucial to be proactive in your research.

Federal and State Tax Credits: Many governments promote the installation of energy-efficient systems by providing tax credits. Visit your country or state's tax department website or consult a tax professional familiar with energy credits.

Utility Company Offers: Utility companies might offer rebates or incentives for customers who reduce their energy footprint. Check your utility's website or contact their customer service for current promotions.

Manufacturer Deals: Sometimes, HVAC system manufacturers provide rebates on specific energy-efficient models. Keep an eye on promotions, especially during seasonal sales or product launches.

How to Avail of These Economic Boons

Keep Updated with Documentation: For tax credits, you'll typically need to submit a form when filing your taxes. Maintain a folder with all purchase receipts, warranty information, and any certificates or documentation stating the energy efficiency of your system.

File On Time: Tax credits and rebates often have strict deadlines. Mark your calendar and ensure you submit all necessary forms and documents punctually.

Professional Consultation: If you're unsure about eligibility or how to claim a rebate or credit, it might be worthwhile to consult a professional. Tax consultants, for instance, can guide you through the maze of paperwork for tax credits.

Stay Updated: Energy efficiency standards and incentives change over time. Regularly check government, utility, and manufacturer websites to stay informed of new opportunities or changes to existing programs.

The Real Economic Impact

It's evident that these incentives can reduce the immediate out-of-pocket costs of installing an energy-efficient system. But there's more to the story:

Reduced Monthly Bills: Beyond the immediate savings from credits and rebates, these systems often lead to lowered monthly utility bills, amplifying the economic benefit over time.

Property Value Increase: Homes equipped with energy-efficient systems may see a rise in property value, as potential buyers recognize the benefits of reduced energy consumption.

Minimized Future Expenses: High-efficiency systems typically last longer and require fewer repairs, leading to further long-term savings.

Harnessing the power of tax credits, rebates, and incentives, homeowners can truly maximize the value they receive from their energy-efficient systems. Beyond the comfort and environmental benefits, the financial gains make the decision to upgrade an astute one. By staying informed and proactive, you can ensure you're not leaving any money on the table when making the switch to more efficient HVAC solutions.

Resale Value and HVAC: Boosting Property Worth with Modern Systems

In today's competitive real estate market, every edge counts. While many homeowners focus on aesthetic improvements like kitchen remodels or landscaped gardens, an often-underestimated aspect of property value is the HVAC system. Modern, efficient HVAC systems can significantly elevate a property's appeal and, consequently, its market value. Let's unpack how an optimized HVAC system can be a game-changer when it's time to sell.

Modern HVAC Systems: More Than Just Temperature Control

Efficiency Is Key: Current HVAC models boast higher energy efficiency than older units. This efficiency translates into reduced energy bills, a significant attraction for potential buyers who are ever more conscious of recurring expenses.

Technological Advances: Today's systems integrate with smart home technologies, allowing homeowners to control temperature settings via smartphones or voice-activated devices. This "smart" functionality appeals to the tech-savvy demographic, a growing segment of homebuyers.

Environmental Impact: A modern HVAC system is not only lighter on the wallet but also on the environment. With eco-awareness at an all-time high, properties boasting eco-friendly features can command higher prices.

Perceived Value: The Buyer's Perspective

The real estate market operates largely on perception. Buyers often base decisions not just on objective factors but also on the emotional appeal of a property.

Move-In Ready Appeal: Buyers often gravitate towards homes requiring minimal immediate work. A state-of-the-art HVAC system signifies a move-in ready home, eliminating the need for immediate replacements or upgrades.

Comfort and Wellbeing: An efficient HVAC system ensures consistent temperature regulation and improved air quality. For buyers, this translates to comfort and a healthier living environment, especially crucial in regions with extreme climates.

Long-term Savings: Forward-thinking buyers factor in long-term costs. While they might be paying a premium for a home with a modern HVAC system, the long-term savings from reduced energy bills can offset the initial investment.

Peace of Mind: Newer HVAC systems come with extended warranties and are less likely to need urgent repairs. This assurance can be a significant selling point, as it offers buyers peace of mind regarding unexpected maintenance costs.

The Economic Implications: Numbers Speak Louder

While the perceived value is vital, the tangible economic benefits of a modern HVAC system are hard to overlook.

Direct Impact on Property Price: Real estate professionals often observe a noticeable price difference between properties with modern HVAC systems and those without. In some instances, homes with updated systems can fetch several thousand dollars more than their counterparts.

Shorter Time on the Market: Properties featuring efficient HVAC systems often sell faster. Quick sales reduce holding costs for sellers (like mortgage, insurance, and utilities) and can sometimes command a higher price due to buyer urgency.

Negotiation Leverage: When a property boasts premium features like a state-of-the-art HVAC system, sellers find themselves in a position of strength during negotiations. This advantage can translate to closer-to-asking price offers.

The Fine Print: Points to Ponder

While the benefits are manifold, homeowners should keep some things in mind:

Documentation Matters: Maintain records of installation, warranties, and maintenance checks. This documentation can serve as proof of the system's condition and efficiency.

Regular Maintenance: While having a modern system is excellent, ensuring it's in optimal working condition is equally crucial. Regular check-ups and timely maintenance can ensure the system impresses during home showings.

Expert Consultation: Considering an upgrade? Speak to an HVAC professional. They can advise on the most suitable system for your property, considering both efficiency and size.

An HVAC system, often viewed as a mundane functional element, can be a property's unsung hero, influencing both its appeal and price. Whether you're a homeowner looking to sell or simply wishing to invest in long-term property value, remember: while curb appeal might draw buyers in, it's the comforts of the interior, primarily dictated by the HVAC system, that seal the deal.

CHAPTER 10

NAVIGATING THE DIGITAL AGE OF HVAC

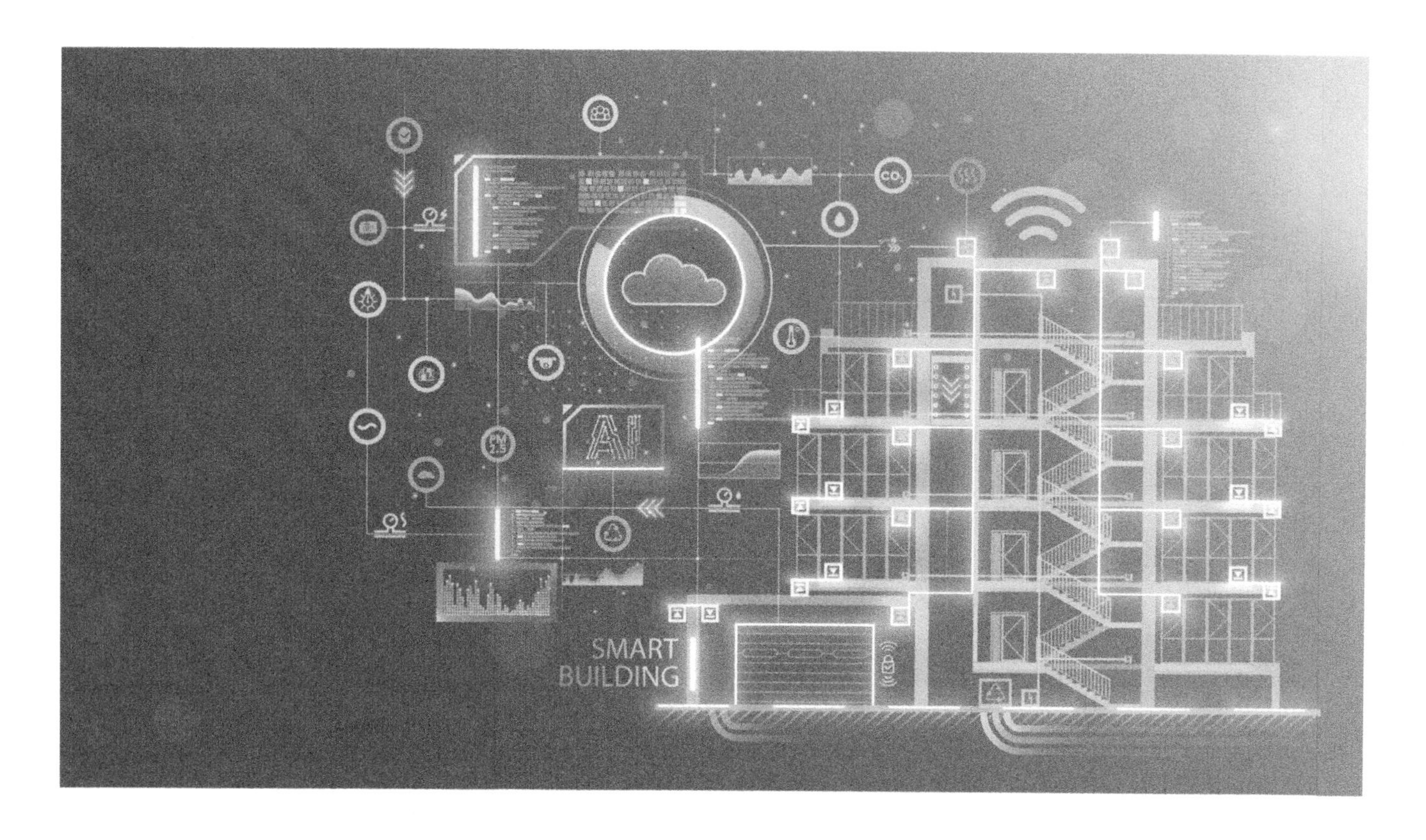

Introduction: HVAC's Spirited Journey into the Digital Frontier

It's hard to imagine the vast chambers of a modern skyscraper or the cozy nooks of our homes without the silent hum of an HVAC system, tirelessly working to keep us comfortable. From humble mechanical roots to the dazzling digital technologies of today, the HVAC industry has witnessed a riveting transformation. In this chapter, we'll journey through HVAC's evolution and explore how digital innovations are reshaping this age-old industry.

HVAC: The Analog Beginnings

Before the razzmatazz of the digital age, HVAC systems were founded on pure mechanics. These were days when regulating room temperature was as simple (or complex, depending on how you look at it) as adjusting a manual dial or opening a vent.

The Era of Mechanical Mastery: The inception of HVAC was marked by mechanical brilliance. Relying solely on the principles of thermodynamics, the pioneers of HVAC developed systems that were manual and quite bulky. Yet, these were groundbreaking, revolutionizing how we interacted with our indoor environments.

The Electromechanical Leap: As homes and businesses became electrified, so did our HVAC systems. This era introduced a fusion of electrical components with mechanical operations. A significant marker of this phase was the introduction of electromechanical thermostats, which brought with them a touch more precision and user-friendliness than their purely mechanical predecessors.

The Digital Usher: Modernizing HVAC

The digital age, characterized by microchips, software, and seamless connectivity, didn't sidestep the HVAC industry. As technology started making waves across sectors, HVAC began its transformation, adopting and integrating these advancements.

Introduction of Electronic Thermostats: The once simple thermostat got a digital makeover. No longer just a dial on the wall, it morphed into a programmable gadget. Homeowners could now set different temperatures for different times, allowing for optimal energy use and enhanced comfort.

Diagnostics and Predictive Maintenance: Digital technology endowed HVAC systems with a newfound intelligence. Systems could now preemptively notify users about potential issues, drastically reducing the downtime caused by unexpected breakdowns. This era marked the transition from reactive problem-solving to proactive maintenance.

Integration with Home Automation: Digital solutions allowed HVAC systems to mesh seamlessly with other home automation tools. With the rise of smart homes, HVAC units began communicating with other devices, optimizing performance based on various inputs, from outdoor weather conditions to the number of occupants in a room.

This foray into the digital age has ensured that modern HVAC systems are not only more efficient and reliable but also more in tune with the needs and preferences of users. As we stand on the brink of further technological advancements, it's exciting to envision where the HVAC industry will journey next.

Digital Controls and Monitoring: A Comprehensive Overview

The technological advancement of HVAC systems has evolved dramatically over recent years. The inception of digital controls and monitoring has marked a significant shift in how homeowners and professionals alike manage heating, ventilation, and air conditioning in residential and commercial spaces. This chapter will delve into the specifics of digital thermostats, applications, and web platforms, underscoring their prominence in today's HVAC world.

Digital Thermostats: More Than Just a Display

Digital thermostats are not merely upgraded versions of their analog predecessors. Their value is rooted deeply in the benefits they provide:

Accurate Temperature Settings: With a digital interface, users can set their desired temperature with precision. This granularity ensures that the indoor environment remains consistently at the intended temperature.

Scheduled Operations: Modern digital thermostats often come equipped with scheduling features. This allows users to program temperature changes around their daily routines, automatically adjusting settings at predefined times.

Applications & Web Platforms: Bridging the Gap

Beyond the physical unit of a thermostat, the integration of HVAC systems with mobile applications and web platforms has further streamlined their management:

Anywhere, Anytime Control: The primary advantage of mobile and web-based platforms is the capability for remote access. Users can change settings, adjust temperatures, or turn systems on or off from virtually anywhere with an internet connection.

Insightful Monitoring: These digital platforms provide a window into system performance. Users can access data such as energy consumption rates, historical temperature settings, and more. This real-time monitoring can be invaluable for identifying patterns and making informed adjustments.

Prompt Notifications: Many advanced digital control systems are designed to provide instant notifications to users. These alerts can range from maintenance reminders, like a filter change, to urgent issues that might require immediate attention.

The Underlying Significance of Digital Controls

Digital controls in HVAC systems aren't just about sophisticated technology. They address practical needs and bring about tangible benefits:

Financial Savings: The precision offered by digital controls often translates to economic benefits. By allowing homeowners to set and adjust temperatures accurately, there's less energy wastage, which can reflect positively on monthly utility bills.

System Longevity: The ability to keep a close eye on HVAC performance through real-time monitoring helps prevent undue strain on the system. By addressing potential issues promptly or making timely adjustments, homeowners can potentially extend the lifespan of their HVAC units.

Sustainability: The global shift towards energy conservation is undeniable. Optimized HVAC operations, facilitated by digital controls, play a part in this. Efficient energy use not only reduces costs but also minimizes the carbon footprint of a household or commercial establishment.

The migration from traditional to digital in the HVAC world isn’t a mere trend. It's a reflection of the broader movement towards integration, efficiency, and sustainability in our daily lives. As HVAC systems continue to evolve, the role of digital controls and monitoring will undoubtedly become even more central.

Integration of Smart Technology in Modern HVAC Systems

As HVAC systems evolve to meet the needs of the contemporary household, smart technology is taking center stage, transforming the ways we heat, cool, and ventilate our spaces. A cornerstone of this transformation is the smart thermostat. Delving deeper, let's understand its place and significance in the modern home.

Smart Thermostats: Revolutionizing Climate Control

The Onset of Wi-Fi-Enabled Thermostats

Traditional thermostats, with their manual dials and limited features, have served households for decades. However, as the world transitioned to a digital era, thermostats too underwent a metamorphosis. The result? Wi-Fi-enabled thermostats, which elevated climate control to new heights of efficiency and precision.

Key Functionalities of Modern Thermostats

Wi-Fi-enabled thermostats are not just about remote temperature adjustments. Their value proposition encompasses a range of functionalities:

Integration with Smart Home Systems: The ability of these thermostats to sync with larger smart home ecosystems is invaluable. Compatible with platforms like Google Assistant, Amazon Alexa, and Apple's HomeKit, they seamlessly weave into the fabric of a connected home.

Learning and Adaptability: Many smart thermostats are now equipped with algorithms that monitor and learn from user behavior. Over a period, they discern patterns in temperature preferences, time-based routines, and more, automating adjustments for optimal comfort.

Real-Time Data Insights: Through embedded sensors and data processing capabilities, these thermostats can offer insights into indoor conditions like humidity and occupancy. This data empowers users to make informed decisions about HVAC settings.

Zonal Climate Control: Going beyond a uniform temperature setting for the entire home, certain smart thermostats provide zonal control. This feature allows users to set varied temperatures for distinct zones or rooms, ensuring customized comfort.

Tangible Benefits of Smart Thermostats

The shift to smart thermostats is not just a tech upgrade; it brings forth a suite of benefits:

Energy Efficiency: One of the most tangible advantages is energy conservation. By autonomously adjusting settings based on user behavior, external weather conditions, and occupancy, smart thermostats ensure minimal energy is wasted. This proactive approach not only promotes sustainability but also reduces utility bills.

Remote Access for Enhanced Convenience: With associated mobile apps, Wi-Fi-enabled thermostats grant users the capability to control HVAC settings from any location. Whether you're in another room or another country, the power to modify your home's climate is just a few taps away on your smartphone.

Anticipatory Heating and Cooling: Leveraging data analytics, some advanced smart thermostats can pre-emptively adjust temperatures. If it's a cold morning, the thermostat can ensure your home is warm as you wake up. Or if you're on your way home, the device can start cooling or heating in anticipation of your arrival.

To encapsulate, the integration of smart technology, exemplified by smart thermostats, marks a significant leap in the evolution of HVAC systems. As households demand more convenience, efficiency, and adaptability, smart HVAC solutions are rapidly becoming the norm rather than the exception. In this context, smart thermostats are a foundational step towards a future of intuitive, responsive, and efficient home climate control.

IoT (Internet of Things) in HVAC: Modernizing Climate Control

HVAC, traditionally perceived as a practical, albeit monolithic domain, is undergoing a paradigm shift. Leading this change is the Internet of Things (IoT), a concept that amalgamates physical devices with internet connectivity to foster intelligent interactions. As HVAC marries IoT, we observe transformative outcomes in energy efficiency, maintenance predictability, and real-time performance optimization.

Sensors: The Sentinels of Modern HVAC

Embedded within the vast landscape of IoT are sensors, minute components that bear a gargantuan responsibility. In HVAC applications, these sensors are the nerve centers, collecting crucial data points to influence decision-making processes.

Temperature and Humidity Sensors: The primary purpose of HVAC is to regulate temperature and maintain comfortable humidity levels. Advanced sensors monitor these parameters with acute precision, ensuring the system operates at optimal efficiency. Moreover, these sensors can adjust HVAC operations based on real-time external weather conditions, preventing overuse and energy wastage.

Occupancy Sensors: These sensors detect human presence in specific zones or rooms. When a room is vacant, there's no need to maintain the same temperature level, and the HVAC system can modulate accordingly. This feature not only conserves energy but also contributes to the longevity of the HVAC system by reducing unnecessary load.

Air Quality Sensors: These detect pollutants, allergens, or particulate matter, adjusting HVAC operations to ensure healthy indoor air. In scenarios where pollutants rise above a certain threshold, the HVAC system can increase ventilation or activate air purifiers.

Predictive Maintenance: Anticipating Before Failing

A significant advantage brought about by IoT in HVAC is the shift from reactive to predictive maintenance. By continuously monitoring system performance and detecting anomalies, IoT-powered HVAC systems can predict potential faults or breakdowns.

Vibration and Acoustic Sensors: These detect unusual vibrations or sounds emanating from HVAC equipment. Uncharacteristic noises often precede component failures, and timely detection can prevent expensive repairs or replacements.

Pressure and Fluid Flow Sensors: Monitoring the flow of coolants or refrigerants is crucial. Any deviations can indicate leaks or blockages, allowing technicians to address issues before they escalate.

Real-time Data: The Key to Performance Optimization

The continuous flow of real-time data is arguably the most potent tool in the IoT-HVAC arsenal. This information, sourced from myriad sensors, offers a panoramic view of system operations.

Energy Consumption Metrics: IoT devices can provide granular data on energy usage, down to individual components. This information can spotlight inefficiencies, guiding homeowners or facility managers in tweaking settings to maximize energy conservation.

Operational Patterns and User Behavior: By analyzing usage patterns, IoT systems can discern user preferences and routines. For instance, if every weekday, the temperature is lowered around 6 PM, the system recognizes this pattern and automates it, ensuring comfort as soon as residents arrive home.

Real-time Diagnostics: IoT doesn't just flag problems; it pinpoints them. If a specific part of the HVAC system is underperforming, the interconnected sensors can identify the issue, facilitating quicker repairs and minimizing downtime.

In this age of digitization, the fusion of IoT with HVAC isn't just a novelty—it's a necessity. As consumers become increasingly conscious about energy conservation, cost savings, and environmental impact, the demand for smart, connected HVAC systems will only intensify. IoT, with its repertoire of sensors and real-time data analytics, stands at the forefront of this HVAC renaissance, championing a future where comfort, efficiency, and sustainability coalesce seamlessly.

Home Automation and HVAC: Making Your Home Truly Smart

Step into the future of home comfort. As technology evolves, our homes have started becoming more intelligent, offering seamless interactions at every corner. At the heart of this smart revolution is home automation. Within this realm, HVAC isn't a mere bystander; it's an active participant, integrating with leading automation platforms like Alexa or Google Home. The result? Elevated comfort, convenience, and control.

Integration: Merging HVAC with Home Automation Ecosystems

When we speak of home automation, we often picture lights that turn off on command or curtains that draw themselves at the appointed hour. However, HVAC has successfully marked its territory within this ecosystem, interfacing with leading home automation systems.

Unified Control: Remember the days when each appliance had its remote or control system? Those days are long gone. With integration, your home automation hub becomes the master control. Adjust your thermostat, toggle the lights, or secure the doors, all from a unified interface, be it an app, a tablet, or a smart speaker.

Synchronized Operations: Imagine your home's thermostat automatically turning down when the windows are open, or the air purifier kicking in when the room's occupancy increases. Integration allows HVAC systems to work in tandem with other smart devices, optimizing energy usage and enhancing comfort.

Speak Up: Commanding Your HVAC

The magic of voice control has breathed life into our homes. With platforms like Alexa or Google Home, the interaction isn't just digital—it's personal.

Voice-activated Comfort: "Alexa, set the thermostat to 72 degrees." With such simple voice commands, you can dictate your comfort. No more fiddling with dials or buttons. Just say the word, and your home adjusts to your preference.

Feedback and Queries: Not just commands, these smart systems can also provide valuable feedback. Ask, "Google, what's the indoor humidity?" and get instant insights into your home's ambient conditions.

Scheduling and Smart Features: Elevating User Convenience

While voice commands are impressive, the true prowess of a smart HVAC lies in its ability to anticipate and adapt. This foresight, coupled with user-defined scheduling and other smart features, transforms the user experience.

Routine-based Scheduling: Are weekdays different from weekends in terms of HVAC needs? Maybe you prefer warmer temperatures when you're snuggling in bed on a lazy Sunday morning. With smart scheduling, your HVAC system can distinguish these routines and adjust accordingly. Define your week, set your preferences, and let the system handle the rest.

Geo-fencing Capabilities: Modern smart HVAC systems can detect your proximity to home. As you approach, the system activates, ensuring your home is at the perfect temperature by the time you step in. Conversely, when you leave, the HVAC goes into an energy-saving mode, cutting down unnecessary costs.

Learning and Adaptation: Some advanced systems go beyond mere scheduling. They learn. By analyzing user behavior over time, these systems predict preferences. So, if you've been lowering the thermostat every night at 10 PM, the system soon starts doing it for you, automating the comfort process.

Integration with Other Smart Features: Home automation is a vast field, encompassing security, entertainment, lighting, and more. Smart HVAC can play along. Visualize a scenario where, as the sun sets, your blinds automatically lower, the room's lights dim to a warm hue, and the temperature nudges up a bit for evening comfort—all orchestrated without a single touch.

The HVAC industry is in the midst of a renaissance. As home automation burgeons, HVAC systems are no longer just about heating or cooling; they're about enhancing lifestyles. By integrating with platforms like Alexa and Google Home, and adopting features such as voice control and smart scheduling, modern HVAC systems are redefining user convenience. In this orchestrated dance of technology and comfort, homeowners stand to gain the most, experiencing a future where their voice, quite literally, shapes their environment.

Zoning Systems and Smart Vents: Optimal Temperature Control for Modern Homes

As we continue our journey into understanding modern HVAC systems, we must delve into two transformative innovations that have redefined home comfort: digital zoning systems and smart vents.

Digital Zoning: Tailored Temperature Management

For a long time, homeowners had little choice but to accept a single temperature setting for their entire household. However, digital zoning systems have ushered in an era where different sections (or zones) of a house can maintain varying temperatures, according to specific needs.

Understanding Zoning: Traditional HVAC systems are designed to treat a home as a single unit, often resulting in uneven temperature distribution. Digital zoning, on the other hand, segments a home into distinct zones, each with its individual thermostat. This allows users to set different temperatures for different areas of the house.

Operational Mechanism: Modern zoning systems utilize advanced thermostats and sensors. These devices continuously monitor temperatures in their designated zones. If a particular zone's temperature deviates from its set point, the system activates, adjusting airflow to bring the zone back to the desired temperature.

Smart Vents: Precision in Airflow Control

While zoning outlines the areas of temperature control, smart vents provide the mechanism to achieve those temperature goals with precision.

Defining Smart Vents: Unlike traditional vents, smart vents are embedded with sensors and have automation capabilities. These vents can adjust their openings based on the temperature needs of a specific zone, either reducing or increasing airflow as necessary.

Achieving Desired Temperatures: By controlling airflow, smart vents can help maintain a room's set temperature. For instance, if a zone has already achieved its desired temperature, the smart vent can limit the airflow, ensuring no excess heating or cooling occurs. This feature not only ensures consistent temperatures but can also contribute to energy savings.

Integration Capabilities: One of the standout features of smart vents is their ability to work in tandem with other smart devices. They can communicate with thermostats, occupancy sensors, and home automation platforms, leading to a cohesive temperature management system.

Zoning and Smart Vents: A Collaborative Approach to Home Comfort

When we discuss digital zoning systems and smart vents, it's crucial to understand that these technologies are complementary.

Digital zoning lays out the temperature blueprint for a house, designating which zones need specific temperature settings. On the other hand, smart vents act on this blueprint, adjusting airflow to ensure each zone achieves and maintains its designated temperature.

Together, they represent a modern approach to HVAC, where temperature control is not only about heating or cooling a home but about doing so with precision and efficiency. Through their combined efforts, homeowners can now enjoy a more customized temperature experience, ensuring comfort in every part of their home.

Artificial Intelligence (AI) & Machine Learning: The Future Frontier of HVAC

When one thinks of HVAC systems, visions of ducts, vents, and thermostats typically flood the mind. Rarely does one picture advanced algorithms and cutting-edge digital intelligence. Yet, as we dive into the modern age, the synergy between HVAC and technology, particularly Artificial Intelligence (AI) and Machine Learning, is undeniable. Let's explore this junction of traditional mechanics and innovative tech.

Predictive Analytics: Elevating Maintenance and Performance

Gone are the simplistic days of HVAC being a mere background function in our homes and offices. Now, with the infusion of AI, these systems can proactively interact, anticipate needs, and respond to potential issues with agility.

The Essence of Predictive Analytics in HVAC:

Predictive analytics revolves around utilizing past data to predict future events or behaviors. In the HVAC world, this means scrutinizing past system performances, identifying patterns, and forecasting potential wear, tear, or malfunctions.

Machine Learning's Magic Touch:

As an intrinsic component of AI, Machine Learning pushes predictive analytics even further. Instead of solely relying on past data, it allows HVAC systems to evolve by "learning" from previous patterns and continuously refining predictions.

Maintenance Transformed:

Predictive capabilities redefine maintenance protocols. Rather than acting reactively once a problem arises, HVAC systems can now forewarn homeowners or technicians about emerging issues. This proactive approach not only prevents inconvenient disruptions but can also spell substantial savings by avoiding complex repairs.

Peak Performance and Efficiency:

Beyond troubleshooting, AI-driven analytics streamline system performance. By discerning data patterns, AI can recommend settings and adjustments that optimize energy consumption. The result? Enhanced efficiency, reduced energy bills, and an extended lifespan for the HVAC unit.

The Limitless Horizon: AI's Role in Future HVAC Innovations

While today's achievements of AI in HVAC are commendable, the horizon is even more promising. The application of AI isn't just a fleeting trend; it's setting the stage for groundbreaking HVAC developments.

Customized Comfort through Adaptive Learning:

Envision an HVAC system that transcends understanding broad climatic patterns. Picture a system that comprehends your unique routines, preferences, and even your moods. With AI, this isn't a futuristic fantasy. HVAC systems can now adapt and "learn" from user behavior, ensuring tailor-made comfort. Whether it's recognizing the time of day you usually feel cold or adjusting temperatures just before you arrive home, adaptive learning is the epitome of personalized luxury.

Enhanced Air Quality Monitoring:

Modern HVAC systems, under AI's guidance, aren't limited to temperature control. They can monitor indoor air quality in real-time, identifying pollutants, allergens, or imbalances in humidity. Armed with this data, the system can make necessary adjustments, ensuring not just comfortable but also healthy air.

Energy Consumption Forecasting:

One of AI's standout features is its ability to predict, and this extends to energy consumption. By analyzing various data points, from weather forecasts to historical energy use, AI can predict future energy needs. This foresight allows users to plan better and potentially capitalize on off-peak energy rates, driving down costs.

Integration with Smart Home Ecosystems:

AI-driven HVAC systems can seamlessly integrate with broader smart home ecosystems. This means your HVAC can communicate with other smart devices, from lights to security systems, creating a harmonious, automated living environment.

Self-repair Protocols:

As AI continues to penetrate the HVAC industry, we can anticipate systems equipped with self-diagnostic and self-repair capabilities. While major malfunctions may still require human intervention, minor glitches could be rectified autonomously by the system, ensuring uninterrupted comfort.

As the realms of AI and HVAC continue to intertwine, the user stands to benefit the most. The combination promises not just advanced mechanics but a user experience that's unparalleled. The future of HVAC is not just smart; it's genius.

Remote Diagnostics and Maintenance: Modernizing HVAC Care

There's a quiet revolution happening in the world of Heating, Ventilation, and Air Conditioning (HVAC). As homeowners, we've been conditioned (pun intended) to brace ourselves for the inevitable moment our systems falter. Traditionally, this meant making an emergency call, waiting anxiously for a technician, and then bracing for the bill. Today, thanks to advancements in digital technology, remote diagnostics and maintenance are reshaping this narrative.

Remote Diagnostics: The Digital Stethoscope for HVAC Systems

In the same way that doctors use stethoscopes to listen to a patient's heartbeat, HVAC technicians now possess their own digital "stethoscope" to gauge the health of a system—without even being on site.

What is Remote Diagnostics?

Quite simply, remote diagnostics involve using advanced software, sensors, and often internet connectivity to monitor, analyze, and diagnose potential issues in an HVAC system from a distant location.

How Does It Work?

Modern HVAC systems are increasingly integrated with a suite of sensors that continuously collect data regarding performance, temperature, air quality, and more. This data can be relayed in real-time to a centralized platform accessible by technicians.

Instant Alerts and Proactive Intervention:

One of the hallmarks of remote diagnostics is the ability to provide immediate notifications when the system detects anomalies. This means before a minor glitch morphs into a major problem, both homeowners and technicians get alerted.

The Compelling Case for Remote Maintenance

The proactive nature of remote diagnostics perfectly segues into its partner-in-crime: remote maintenance.

Predictive Maintenance—A Game Changer:

Instead of adhering to a traditional schedule or waiting for something to break down, predictive maintenance relies on the data gleaned from remote diagnostics. It allows technicians to intervene when the system signals the onset of a potential issue. Think of it as giving your HVAC a "check-up" precisely when it needs one.

Cost Efficiency:

Predictive maintenance spells significant cost savings. By addressing potential problems early on, homeowners can avoid the steep costs associated with major breakdowns. Moreover, by maintaining the system at its optimal performance level, energy consumption stays efficient, reflecting favorably on utility bills.

Minimizing In-Person Visits:

While there are instances where a technician's physical presence is irreplaceable, remote maintenance reduces the frequency of these visits. For routine checks or minor adjustments, a technician can often make tweaks remotely. This is especially beneficial in our increasingly digital age where convenience is paramount and minimizing disruptions is always appreciated.

Predictive vs. Reactive Maintenance: Why Being Proactive Pays Off

The move from reactive to predictive maintenance marks a pivotal shift in how we approach HVAC care. But why is this shift so valuable?

Extended HVAC Lifespan:

Consistent, proactive care ensures that an HVAC system runs smoothly, experiencing less wear and tear. This not only ensures efficient performance but can also extend the system's lifespan.

Avoiding the Domino Effect:

In HVAC systems, one malfunctioning component can trigger a cascade of issues. Predictive maintenance, by addressing issues at their nascent stage, can prevent this domino effect, saving homeowners from more extensive (and expensive) repairs.

Peace of Mind:

There's an intangible yet invaluable benefit to predictive maintenance—peace of mind. Knowing that your HVAC is continuously monitored and that professionals are ready to intervene at the first hint of trouble provides homeowners with a comforting reassurance.

The world of HVAC is no longer just about heating and cooling spaces. It's about harnessing technology to deliver convenience, efficiency, and proactive care. With remote diagnostics and maintenance, the future of HVAC isn't just about responding to problems—it's about anticipating them. As homeowners and industry professionals embrace this new paradigm, the traditional woes of HVAC care are set to become a thing of the past. Welcome to the era where your HVAC not only "talks" but also "listens" to its own needs.

Energy Analysis Tools: Deciphering the Digital Power Print of HVAC Systems

When we discuss HVAC systems, our conversations often revolve around temperature settings, seasonal adjustments, and that familiar hum signaling its operation. However, beneath the tangible world of breezy airflows and thermostat settings lies a matrix of data, charts, and numbers that hold the key to unlocking HVAC efficiency. Welcome to the realm of energy analysis tools.

Energy Analysis Tools: A Brief Introduction

At the core, energy analysis tools are sophisticated software and digital platforms that tap into the vast data reservoir generated by modern HVAC systems. These tools, like detectives equipped with magnifying glasses, scrutinize this data to reveal insights about energy consumption patterns. It's akin to having a financial advisor for your energy bills – guiding where to save and where to invest.

Peeling Back the Layers of Consumption with Digital Tools

Real-Time Monitoring:

The first step to understanding your energy consumption is knowing what's happening in real-time. Several digital tools provide instantaneous readouts of your system's energy usage. This offers a clear snapshot of how different components or zones in your HVAC system consume energy at any given moment.

Historical Data Review:

Beyond the 'now,' these tools are also time travelers of sorts. They can pull up historical data, allowing homeowners to compare energy usage across different periods. Want to see how much energy your system consumed last winter compared to this summer? No problem!

Energy Consumption Breakdown:

Have you ever looked at your energy bill and wondered, "Where did all that energy go?" Digital energy analysis tools can provide a granular breakdown. This might mean seeing how much energy goes into heating versus cooling or even the energy consumption of specific rooms or zones.

Translating Data Into Tangible Savings: The 'Recommendation Engine'

Having a pool of data is one thing; understanding what to do with it is another. This is where the recommendation functionalities of energy analysis tools shine.

Optimizing Temperature Settings:

One of the most straightforward recommendations you might receive is about your thermostat settings. Sometimes, a slight adjustment, maybe a degree or two, can lead to substantial savings without compromising comfort.

Scheduling Recommendations:

For those using programmable thermostats, energy analysis tools can suggest the most energy-efficient schedules. This could mean adjusting the timings for when your HVAC kicks in during the morning or optimizing its operation for when you're typically away.

Zone-Specific Insights:

Not all rooms are created equal, and your living room might have different HVAC needs than your bedroom. Energy analysis tools can offer zone-specific insights, guiding homeowners on where they might need to tweak settings or even implement additional insulation.

Maintenance Alerts:

These tools can also predict when specific components might be due for maintenance. A filter that needs changing or a component that's working harder (and consuming more energy) than it should can be flagged, ensuring the system operates at peak efficiency.

Empowering Homeowners in the Digital Age

The beauty of these digital energy analysis tools is not just in their ability to reveal patterns but in their power to democratize information. In an era gone by, homeowners would be reliant on technicians or experts to provide insights into their HVAC system's efficiency. Now, with a few taps on a screen, they have this knowledge at their fingertips.

By utilizing these tools, homeowners are better positioned to make informed decisions. Whether it's making an adjustment to the thermostat, scheduling a maintenance check, or investing in system upgrades, these decisions are data-backed, ensuring they're both energy-efficient and cost-effective.

Today's HVAC systems, with their array of sensors, smart components, and digital interfaces, are more than just machines—they're repositories of valuable data. Energy analysis tools are the bridge that connects homeowners to this data, translating it into actionable insights. As we move deeper into the digital age, these tools are not just optional add-ons but essential companions for anyone serious about optimizing their HVAC system's performance and efficiency.

Challenges and Concerns: Navigating the Flip Side of HVAC's Digital Age

In a world of rapidly advancing technology, the HVAC domain hasn't remained a bystander. As we've previously journeyed through the innovative strides of digital tools and automation in HVAC, there's no doubt that we're experiencing a revolutionary shift in how we manage and understand our home's heating and cooling. But, as with all revolutionary advancements, there are hurdles along the way. Let's dive into some of these challenges and chart out informed paths forward.

Cybersecurity Risks: When Your HVAC Talks to the Web

Imagine this: You've just set up your brand-new smart thermostat. It's sleek, touch-sensitive, and Wi-Fi-enabled. You can adjust your home's temperature from the other side of the world if you want. But herein lies a concern - if you can access your HVAC remotely, what's stopping a malicious entity from doing the same?

The Threat Landscape:

With the integration of IoT in HVAC, the systems aren't just communicating with you but potentially with a network of other devices. This connectivity opens doors to cyber threats. Unauthorized access could lead to someone maliciously cranking up your heat or shutting down your AC, leading to discomfort, or worse, damage.

Safeguarding the Sanctum:

DIY Tip: Always change default passwords on your smart devices. This simple step can deter a vast majority of opportunistic hacks. Additionally, regularly updating device firmware ensures you benefit from the manufacturer's latest security patches.

The Double-Edged Sword of Data:

The extensive data your HVAC system gathers for energy analysis and predictive maintenance is invaluable. But, this same data, if mismanaged or accessed unlawfully, can be a privacy concern. Ensure you're aware of how your data is stored, whether it's encrypted, and what the manufacturer's data policies entail.

Continuous Updates: Blessing or Bane?

In the digital realm, the only constant is change. Software updates have become a part of our life - from smartphones to apps, and now, our HVAC systems.

The Why of Updates:

Updates aren't whimsical changes made by developers. They're vital. They could be patching security vulnerabilities, optimizing performance, or introducing new features. In the context of HVAC, updates can translate to a system that runs more efficiently, saves you more money, and lasts longer.

The Challenge of Keeping Up:

However, the frequent need to update can pose challenges:

Downtime: Even if it's just a few minutes, during an update, certain functionalities of your HVAC system may be temporarily unavailable. This is rarely a significant concern, but imagine running an update during a particularly chilly night or scorching day.

Incompatibility: An update meant for one component of the HVAC system might not play well with another older component. It's essential to ensure all devices within the system are compatible post-update.

DIY Tip: Schedule updates during times when HVAC demand is typically low in your household. This might mean during work hours or late at night. This way, you ensure minimal disruption.

The Learning Curve:

An updated system can sometimes mean a changed user interface or new features to get acquainted with. While this promises enhanced user experience in the long run, the immediate aftermath of an update can be a bit befuddling for users.

DIY Tip: Most manufacturers release user guides or update logs detailing the changes. Take a few moments to go through these. They can drastically shorten the time you spend figuring out the 'new' in your system.

Navigating the Path Ahead

While these challenges might seem daunting, especially if you're just venturing into the digital HVAC landscape, they're surmountable. The key lies in being informed and proactive.

Be Knowledge-Empowered: Just as you'd read the user manual of a new gadget, invest time in understanding the digital nuances of your HVAC system. Familiarize yourself with the manufacturer's update cycles, support channels, and cybersecurity recommendations.

Seek Expert Advice: While the DIY spirit is commendable, sometimes it's beneficial to consult experts, especially when setting up intricate HVAC digital systems. They can offer insights, best practices, and even customize configurations to align with your specific needs.

Adopt a Proactive Stance: Instead of reacting to issues, adopt a proactive stance. Schedule regular system checks, keep abreast of software update notifications, and consider joining forums or user groups focused on HVAC digital trends.

While the digital age of HVAC presents challenges, it also offers immense potential - from energy savings and predictive maintenance to enhanced user comfort. As homeowners and enthusiasts, by addressing these challenges head-on, we're not just navigating this new age; we're pioneering the next phase of HVAC's digital evolution.

CHAPTER 11
THE FUTURE OF HVAC

Trends in Miniaturization: Big Performance in a Small Package

The HVAC world, much like our smartphones, computers, and almost every piece of technology we own, is heading in a direction that few might have predicted a few decades ago: the race to become smaller. As urban settings evolve and the demand for efficient solutions in constricted spaces grows, the HVAC industry is innovating in ways that mirror the miniaturization trend observed in many tech sectors.

Compact Systems: The Art of Size Reduction

Going small isn't merely about downsizing the physical unit. It's a holistic approach that ensures performance doesn't take a backseat. Over the years, HVAC systems have become more streamlined, with many internal components undergoing miniaturization. This intricate evolution involves:

Advanced Materials: With the introduction of newer materials that can conduct heat better and stand up to wear and tear, HVAC components can be crafted smaller without compromising efficiency.

Innovative Design: The architecture inside HVAC units has become more sophisticated, allowing components to be closely packed without overheating or interference.

High-Efficiency Compressors: Modern compressors are marvels of efficiency, performing at levels that older, bulkier models couldn't achieve.

DIY Tip: If you're considering an HVAC upgrade, check for the latest compact models in the market. Not only are they space-savers, but many newer models also boast of energy-saving features.

Urban Settings: Where Every Inch Counts

As urban areas become increasingly populated, space is a luxury. Every square foot matters, and bulky HVAC units are a luxury few can afford, both in terms of space and aesthetics. Here's where miniaturized HVAC systems come into play:

Apartment Living: Modern apartments, especially in high-rise settings, often have limitations when it comes to outdoor unit placements. Compact systems can fit into restricted spaces without being an eyesore.

Architectural Aesthetics: For property developers and homeowners who emphasize design, miniaturized systems are a godsend. They're easier to conceal, ensuring that the HVAC system doesn't detract from the property's aesthetic appeal.

Balcony and Rooftop Units: As rooftop gardens and balcony spaces become popular relaxation zones, having a bulky HVAC unit can be a downer. Compact units integrate seamlessly, allowing residents to make the most of these outdoor spaces.

DIY Tip: For those living in apartments, always consult with your property manager or homeowner's association before making modifications or installations to ensure compliance with building regulations.

Retrofitting Older Structures: Breathe New Life with Small Systems

One of the challenges with older buildings, be it homes or commercial spaces, is integrating modern amenities without major renovations. The brickwork might be sturdy, and the wooden beams might ooze vintage charm, but they weren't designed for today's HVAC units. This is where miniaturization becomes a game-changer:

Less Invasive Installation: Traditional HVAC systems might require significant modifications - bigger holes, reinforcements, and more. Compact systems, on the other hand, can often be installed with minimal structural changes.

Preserving Historical Integrity: For heritage properties, maintaining the original look is crucial. Smaller HVAC units can be discretely placed, ensuring that the building's historical essence remains untainted.

Optimized for Efficiency: Older buildings might not have the insulation standards of modern constructions. Compact HVAC systems, especially those designed for retrofitting, are often optimized for higher efficiency, ensuring that they perform effectively even in less-than-ideal conditions.

DIY Tip: If retrofitting an older property, consider working with an HVAC consultant who specializes in vintage or historic homes. Their expertise can guide you to solutions that are both efficient and respectful of the property's character.

The move towards miniaturization in the HVAC sector is not merely a nod to modern design preferences but a testament to the industry's commitment to innovation and adaptability. As urban landscapes change and the value of space increases, the industry's pivot to compact yet powerful systems ensures that comfort, efficiency, and aesthetics coexist. For DIY enthusiasts, homeowners, and property developers alike, navigating the world of miniaturized HVAC systems promises exciting possibilities for the future.

Innovations in Sustainable HVAC Solutions:

Geothermal Heat Pumps: Earth's Gift to Comfort

In our continuous pursuit of eco-friendly and efficient HVAC solutions, we're often inspired by nature. One such inspiration is the geothermal heat pump—a technology that leverages the Earth's consistent temperature as a medium for heating and cooling. Let's unravel this groundbreaking innovation.

Tapping into Earth's Natural Reservoir

For those unfamiliar with the term, "geothermal" stems from Greek words meaning "Earth" and "heat". At its core, geothermal heat pumps, or GHPs, operate on a simple principle: beneath the Earth's surface, the temperature remains relatively constant, unaffected by the changing seasons. This consistent temperature provides a stable, renewable source of heat in the winter and a sink for heat in the summer.

The operational mechanics are fairly straightforward:

1. **Ground Loop System**: A GHP system comprises pipes buried in the ground, either vertically or horizontally. These pipes contain a fluid, usually a mix of water and antifreeze.
2. **Heat Exchange Process**: During winters, the fluid in the pipes absorbs heat from the ground. The heat pump then extracts this heat to warm up the building. In the summer, the process is reversed: the fluid absorbs heat from the building and dissipates it into the ground, providing a cooling effect.

DIY Tip: Curious about the depths? Horizontal systems are generally buried about four to six feet deep, where the temperature remains stable. Vertical systems, however, can go several hundred feet down, ideal for properties with limited land space.

Eco-friendly Savings: The Dual Benefit

Opting for a GHP isn't just about embracing the latest technology; it's about making an environmentally conscious decision that also happens to be light on your wallet. Here's how:

1. **Efficiency Masterclass**: GHPs are incredibly efficient. For every unit of electricity used to operate the system, GHPs can provide up to five units of heating or cooling.

This is substantially higher than traditional HVAC systems, which might, at best, offer an efficiency ratio of 2:1.

2. **Cost Savings**: The initial installation cost of a GHP system might be higher than conventional HVAC systems. However, the reduction in operational costs can offset this within a few years. Some homeowners report savings of up to 50% on their heating and 25% on their cooling bills.
3. **Eco-Friendly Operation**: Unlike combustion-based heating systems, GHPs produce no carbon monoxide or other harmful greenhouse gases. This means you're not only saving money but also significantly reducing your carbon footprint.

DIY Tip: To maximize savings, ensure that your building is well-insulated. This will reduce the workload on the GHP system, further increasing its lifespan and operational efficiency.

Maintenance and Longevity: A Bonus Advantage

While cost savings and eco-friendliness are the primary attractions, GHP systems also boast of low maintenance requirements. The indoor components have a lifespan similar to conventional HVAC systems, typically 20-25 years. However, the ground loops, being protected underground, can last upwards of 50 years, making them a lasting investment.

DIY Tip: Even with minimal maintenance, it's a good practice to have the system inspected every few years. Regular checks ensure that the fluid levels are optimal and there are no leaks or other issues.

The march towards sustainable HVAC solutions is an exciting journey, with Geothermal Heat Pumps leading the charge. Harnessing the Earth's natural temperature stability, these systems epitomize efficiency and eco-consciousness. For homeowners and property developers eyeing the future, integrating GHPs is more than just an HVAC decision—it's a commitment to sustainable living and future-focused adaptability.

Solar-powered HVAC Systems: Let the Sun Power Your Comfort

In a world increasingly conscious of its carbon footprint, every sunbeam offers an opportunity. From generating electricity to heating water, solar power has firmly established itself as a cornerstone of renewable energy. But what about using solar energy for heating and cooling our homes and offices? Enter solar-powered HVAC systems, a synthesis of sustainable practices and modern comforts.

Solar-Powered HVAC: How Does It Work?

At its essence, a solar-powered HVAC system uses solar panels, typically photovoltaic (PV) cells, to convert sunlight into electricity. This electricity then powers the HVAC system, providing heating, cooling, and ventilation as required.

1. **Solar Panels**: These are the system's workhorses. They capture sunlight and convert it into direct current (DC) electricity.
2. **Inverters**: Once the DC is generated, inverters come into play. They convert this DC into alternating current (AC), which is the type of electricity most HVAC systems use.
3. **HVAC System**: With electricity generated, the HVAC system operates similarly to traditional setups, providing the necessary heating or cooling.

DIY Tip: Curious about placement? While most solar panels are roof-mounted, they can also be ground-mounted if your property has ample space. The key is to place them where they'll receive maximum sunlight.

Bridging the Gap: Dealing with Intermittency and Storage

Harnessing solar power for HVAC is thrilling, but the sun isn't always shining. Nightfall and cloudy days introduce intermittency, posing a challenge for solar-driven systems. So, how do we ensure consistent HVAC performance?

1. **Battery Storage Systems**: These are the saviors when the sun takes a break. When solar panels produce excess electricity (think of a bright sunny day), instead of wasting it, this excess is stored in batteries. During off-peak solar hours, the HVAC system can then draw from these batteries.

2. **Hybrid Systems**: Some solar HVAC systems are hybrids. When solar energy is insufficient, these systems can switch to the grid or alternative energy sources. This ensures continuous operation without hiccups.

3. **Thermal Storage**: For systems focusing on solar heating, solar thermal storage is a game-changer. Here, excess thermal energy heats a storage medium (like water or special fluids) during the day, which can then be used at night or during cloudy days.

DIY Tip: If you're considering a solar HVAC setup, evaluate your location's solar potential. Regions with more sunny days can rely more heavily on solar, while cloudier regions might benefit from hybrid systems.

A Ray of Benefits: Beyond Just Sustainability

Solar-powered HVAC systems are more than just 'green'. Here's what makes them shine:

1. **Economic Incentives**: Over time, the savings on electricity bills can be substantial. Additionally, various government incentives and tax breaks for solar installations can offset the initial investment.

2. **Reduced Carbon Footprint**: Solar power is clean. By transitioning, you're contributing to reduced greenhouse gas emissions and a healthier planet.

3. **Independence from Grid**: Power outages? Fluctuating electricity prices? With a solar-powered HVAC system, you can reduce reliance on the grid, ensuring consistent comfort.

DIY Tip: Maintenance is key. Clean solar panels capture more sunlight. Regularly check for dirt, leaves, or snow that might obstruct your panels.

Tapping into the sun's abundant energy for our HVAC needs is more than just innovation—it's a reflection of our evolving relationship with nature and technology. Solar-powered HVAC systems symbolize a future where sustainability and comfort go hand in hand. As homeowners, businesses, and industry professionals, embracing this change is not just about harnessing the sun, but about illuminating a path for generations to come.

Ice-powered Air Conditioning: Harnessing Ice for Sustainable Cooling

Ice-powered air conditioning systems signify a progressive approach in the realm of sustainable HVAC. Although the concept of utilizing ice for cooling is historically recognized, technological advancements have positioned it at the forefront of eco-friendly HVAC innovations.

How Does Ice-Powered Air Conditioning Work?

The mechanism behind ice-powered air conditioning is straightforward:

1. **Ice Creation Phase**: Primarily during off-peak hours, often nighttime, water circulates around specific coils within an ice-making unit. The primary objective during this phase is to freeze this water, leading to the formation of substantial ice blocks.

2. **Cooling through Melting Phase**: As day breaks and temperatures rise, the system shifts its function. The previously formed ice starts melting. As a basic scientific principle, the melting process of ice absorbs heat from its surroundings. In an HVAC context, this principle is harnessed wherein the melting ice cools the circulating refrigerant or water. This cooled medium is subsequently pumped throughout the building, efficiently extracting heat from indoor spaces and facilitating the desired cooling effect.

Advantages of Ice-Powered Air Conditioning Systems

1. **Cost Efficiency during Peak Hours**: One of the system's salient features is its ability to manage energy consumption adeptly. By focusing the energy-intensive process of freezing water during nighttime, when electricity rates are generally lower, it achieves

cooling during the day without significant energy use. This transition to nighttime energy consumption translates to appreciable cost savings.

2. **Operational Continuity during Power Interruptions**: Though no system can claim complete immunity to power disruptions, ice-powered air conditioning systems have an inherent advantage. Given that the ice formation process occurs during the night, the stored ice can continue its melting (and thus, cooling) process for several hours even in the event of a daytime power outage.

3. **Eco-friendly Operation**: Using electricity predominantly during off-peak hours, which often aligns with higher contributions from renewable energy sources, can lead to a reduced carbon footprint. Furthermore, by reducing energy use during peak periods, these systems mitigate the grid's strain, diminishing the necessity for auxiliary and often less eco-friendly power generation.

4. **Enhanced System Longevity**: A recurrent issue with conventional air conditioning units is the pronounced strain they experience during daytime peak temperatures. This strain can lead to accelerated wear and tear. In contrast, ice-powered systems, which significantly reduce daytime operational strain, can have extended operational lifespans.

Applicability of Ice-Powered Air Conditioning

The efficiency of ice-powered air conditioning is influenced by certain external factors. Areas that experience considerable differences in day and night temperatures can derive the maximum benefit from these systems. Moreover, for regions where electricity pricing varies based on demand (such as time-of-use pricing structures), the economic advantages of ice-powered air conditioning become even more pronounced.

In summary, ice-powered air conditioning stands as a testament to the HVAC industry's commitment to integrating sustainability with operational efficiency. As technology continues its onward march, such innovations are pivotal in aligning comfort with ecological responsibility.

Thermally Driven Air Conditioning: The Future is Warm and Cool!

Air conditioning has traditionally been associated with electrical power. However, modern innovations in HVAC technology are reshaping these norms, driving efficiency and sustainability through non-conventional avenues. Among these pioneering solutions is thermally driven air conditioning, which paradoxically uses heat to achieve coolness. Let's delve deeper into this revolutionary methodology.

Thermally Driven Air Conditioning: The Science Behind the Coolness

At its core, thermally driven air conditioning operates on the principles of absorption refrigeration. Instead of relying on an electrically powered compressor to circulate the refrigerant, this technology uses heat sources, like solar energy or natural gas, to trigger a heat exchange process. This, in turn, drives the cooling mechanism.

1. Harnessing Solar Heat: In the case of solar-driven systems, solar collectors, often in the form of panels, capture and convert sunlight into heat. This captured heat is then employed to boil a special refrigerant out of an absorptive solution. As this refrigerant vaporizes and condenses, it induces a cooling effect, similar to the process within traditional air conditioners but without the electrical demands of compressors.

2. Tapping into Natural Gas: For settings where consistent solar input might be a challenge, natural gas serves as an alternate heat source. Using a burner, the natural gas is combusted to generate the requisite heat for driving the absorption process, ensuring uninterrupted operation even during overcast days or nighttime.

Desiccants: The Unsung Heroes of Enhanced Efficiency

When discussing thermally driven air conditioning, it's impossible to overlook the pivotal role of desiccants. These are materials known for their inherent ability to absorb moisture. Integrating desiccant solutions into a thermally driven air conditioning system can achieve a dual objective.

1. Humidity Control: Desiccants, due to their moisture-absorbing attributes, can effectively control indoor humidity levels. By extracting excess moisture from the ambient air, they ensure that the cooled environment remains comfortable and devoid of the stickiness often associated with high humidity settings.

2. Boosting Cooling Efficiency: A lesser-known fact is that drier air can be cooled more efficiently. By integrating a desiccant system, which pre-treats and reduces the humidity of incoming air, the overall cooling mechanism experiences less strain and consequently, can achieve desired temperatures more rapidly and with less energy.

Key Benefits of Thermally Driven Air Conditioning

1. Reduced Electrical Demands: Given that the primary driving force is heat, the system's electricity consumption is minimal, often restricted to powering fans or minor ancillary components.

2. Sustainable Energy Utilization: When utilizing solar heat as a power source, the system taps into an inexhaustible and eco-friendly energy reserve. This not only reduces carbon footprints but also aligns with broader global objectives of renewable energy adoption.

3. Cost Efficiency: By bypassing peak electricity rates, especially in areas with time-of-use pricing structures, homeowners and commercial entities can realize significant savings.

4. Adaptable Operations: The capability to switch between solar heat and natural gas ensures that the cooling system remains operational regardless of external factors like weather variations.

A Future-Ready Solution

Thermally driven air conditioning is not just a novelty; it's a testament to the HVAC industry's commitment to innovative, sustainable, and efficient solutions. As urban settlements grow and energy demands surge, solutions that harness renewable energy sources without compromising on operational efficiency become not just desirable but indispensable.

This technology, while being future-centric, also nostalgically harks back to ancient times when civilizations used basic thermal principles to manage indoor climates. What was once a rudimentary understanding of heat exchange has now been refined, optimized, and integrated into modern structures, illustrating that sometimes, looking back can indeed propel us forward.

In the grand spectrum of HVAC evolution, thermally driven air conditioning is both a milestone and a beacon, illuminating the path toward a cooler, sustainable future.

Smart Windows and Building Designs: The Unsung Heroes of HVAC Efficiency

In our quest for a comfortable indoor environment, the heart of our focus often lies with HVAC systems. But as the narrative on sustainable energy and efficiency evolves, we're beginning to recognize the paramount importance of supporting players in this grand theatre of indoor climate control. Two such unsung heroes are smart windows and strategic building designs.

Smart Windows: More than Just a Pane

The modern window is no longer a mere transparent barrier between the indoors and the outside world. With the advent of smart window technologies, it's actively contributing to the maintenance of ideal indoor temperatures.

1. Regulating Infrared Light Entry: One of the critical innovations in window technology is the ability to control the passage of infrared (IR) light. Why is this important? IR is responsible for the heat we feel from sunlight. By regulating its entry, smart windows directly influence indoor temperature.

- **Electrochromic Windows**: These windows use a small electrical charge to shift their opacity, thereby controlling the amount of sunlight and heat that penetrates indoors. With just a flick of a switch or even automatically, they can tint to reduce incoming IR light on a hot day or become transparent on cooler ones.

- **Thermochromic and Photochromic Windows**: These smart windows respond to heat or light, respectively. They automatically adjust their transparency levels depending on external conditions, ensuring a more consistent indoor temperature.

Building Orientation and Design: Strategic Comfort

Before delving into the intricacies of HVAC systems, there's merit in taking a step back and evaluating the structure that houses it. The design and orientation of a building play a pivotal role in dictating its thermal performance.

1. Building Orientation: At its simplest, the orientation refers to how a building is positioned in relation to the sun's path. By strategically orienting a structure:

- **Optimal Sunlight Absorption**: In colder climates, a building can be oriented to maximize sunlight absorption during winter months, reducing the need for heating. Conversely, in warmer regions, minimizing direct sunlight can cut down on cooling demands.

- **Natural Ventilation**: A building's orientation can also harness prevailing winds for natural ventilation, promoting passive cooling and reducing reliance on HVAC systems.

2. Architectural Design Elements: Beyond mere orientation, several design techniques can significantly enhance HVAC efficiency:

- **Thermal Mass**: Materials like brick or concrete can store heat during the day and release it at night, assisting in temperature regulation.

- **Green Roofs**: Planting vegetation on rooftops isn't just aesthetically pleasing. It offers an insulation layer, reducing the building's heat absorption.

- **Shading Solutions**: Architectural elements like eaves, pergolas, and awnings can shield windows from direct sunlight, mitigating heat gain.

- **Natural Insulation**: Using insulating materials in walls, roofs, and floors can dramatically reduce the need for active heating or cooling. Furthermore, innovations like double or triple-glazed windows add an extra layer of thermal protection.

The Symbiotic Relationship of Windows, Design, and HVAC

It's tempting to view HVAC systems as independent entities, solely responsible for a building's climate control. However, as we delve deeper into the realm of sustainable solutions and efficiency, it becomes abundantly clear that a more holistic approach is not just beneficial—it's imperative.

Smart windows, with their ability to dynamically regulate light and heat entry, actively contribute to reducing the strain on HVAC units. They preemptively tackle potential temperature fluctuations, ensuring that air conditioning systems aren't constantly playing catch-up.

Similarly, a building's orientation and design act as the first line of defense against unwanted temperature swings. By optimizing these elements, we can significantly reduce the operational demands placed on HVAC systems, prolonging their lifespan, reducing energy consumption, and ensuring a more consistent indoor climate.

In essence, the journey to HVAC efficiency is a collaborative dance, where each element—from smart windows to strategic design—plays its part in harmony. The result? A comfortable interior, reduced energy bills, and a lighter carbon footprint. The future of HVAC, it seems, isn't just about the systems themselves, but the environment in which they operate.

Eco-friendly Refrigerants: Navigating the Green Path in HVAC

The world of HVAC, like many other industries, isn't just about technological advancements; it's also deeply intertwined with environmental concerns. At the epicenter of these environmental debates lie refrigerants, the unassuming substances responsible for the heating and cooling cycles of our HVAC systems. As the world wakes up to the pressing issue of global warming, the call for greener refrigerants is louder than ever.

Refrigerants and Their Impact on Our Planet

To understand why there's a push towards eco-friendly refrigerants, we must first grasp the environmental cost of traditional refrigerants. Many older HVAC systems use hydrofluorocarbons (HFCs), chlorofluorocarbons (CFCs), or hydrochlorofluorocarbons (HCFCs). While these refrigerants are excellent at their job of heat transfer, they have one major flaw: a staggeringly high Global Warming Potential (GWP). In other words, if these refrigerants leak into the atmosphere, they trap heat at a much higher rate than carbon dioxide, significantly accelerating global warming.

The Green Alternatives: Paving the Way for Sustainable Cooling

With the environmental stakes so high, the HVAC industry has been on a quest to find sustainable refrigerant alternatives. Let's dive into some of the most promising ones:

1. CO2 (Carbon Dioxide):

- **How It Works**: It might surprise many to learn that CO2, often vilified as a greenhouse gas, is now being championed as a refrigerant. Used in a transcritical cycle, CO2 can act as an effective cooling agent in HVAC systems.
- **Why It's Green**: As a refrigerant, CO2 has a GWP of 1, which is negligible compared to the thousands attributed to traditional refrigerants. Plus, it's non-toxic and non-flammable, marking it as a safe option.

2. Ammonia (NH3):

- **How It Works**: Ammonia isn't new to the refrigeration game. It's been used for over a century, particularly in industrial settings. With zero ozone depletion potential and a GWP of less than 1, it's easy to see why.
- **Why It's Green**: Apart from its negligible GWP, ammonia's efficiency means systems need less energy to achieve the desired temperature, resulting in reduced carbon emissions from power sources.

3. Hydrocarbons (Propane, Isobutane):

- **How It Works**: Hydrocarbons are organic compounds that have proven to be effective refrigerants in smaller HVAC systems and appliances.
- **Why It's Green**: Hydrocarbons boast a GWP that's below 5. Their efficiency also reduces energy consumption. However, they're flammable, so it's essential to handle them with care and ensure systems are leak-tight.

Embracing the Shift: The Challenge and Triumph

Transitioning to eco-friendly refrigerants isn't merely a matter of swapping out one substance for another. Systems designed for traditional refrigerants might not be compatible with their greener counterparts, necessitating retrofits or entirely new installations. This can represent significant initial costs and pose challenges in terms of widespread adoption.

However, the long-term benefits of these eco-friendly refrigerants are undeniable. Not only do they substantially reduce the carbon footprint of HVAC operations, but many of these green alternatives also offer improved efficiency, which can translate to cost savings in operational expenses over time.

The Role of Policy and Industry

Recognizing the immense environmental benefits, many governments worldwide have started to legislate the phase-out of high-GWP refrigerants. This regulatory push, combined with a genuine industry desire to adopt sustainable practices, has significantly accelerated the transition to eco-friendly alternatives.

Companies are investing heavily in research and development, not only to refine the use of substances like CO2, ammonia, and hydrocarbons as refrigerants but also to unearth new, sustainable solutions. Collaboration between policymakers, industry leaders, and researchers will be the cornerstone of the next era of HVAC – an era defined by efficiency and sustainability.

In our HVAC journey, it's important to remember that every component, down to the refrigerants we use, plays a role in shaping the industry's environmental impact. As we continue to innovate, refine, and adopt these green alternatives, we're not just ensuring comfortable interiors; we're taking a decisive step towards a cooler, sustainable planet.

Air Purification and Enhanced Filtration: Innovations in HVAC Solutions

The Need for Clean Indoor Air

The increasing urbanization and industrialization have led to concerns regarding indoor air quality, especially in densely populated urban areas. While HVAC systems have traditionally focused on heating and cooling functions, the demand for cleaner indoor air has necessitated a pivot towards incorporating air purification mechanisms. The quality of indoor air is no longer just about comfort—it's about health.

Advancements in Filter Technology

The backbone of any HVAC system aiming to purify air is its filtration system. The efficacy of these filters in trapping contaminants determines the quality of the circulated air.

1. HEPA Filters: HEPA (High-Efficiency Particulate Air) filters represent the gold standard in air filtration. Designed to capture at least 99.97% of particles 0.3 microns in diameter, HEPA filters effectively eliminate common allergens such as pollen, mold spores, and dust mites.

2. Activated Carbon Filters: Addressing a different set of contaminants, activated carbon filters specialize in removing gases and odors. They contain a vast surface area which adsorbs these pollutants, making them particularly useful in environments where odor elimination is crucial.

3. Electrostatic Filters: A step up from traditional mechanical filters, electrostatic filters utilize an electrical charge to attract and capture airborne particles. They are also reusable, providing both an economical and environmentally-friendly option for long-term use.

Embracing the Power of UV Light

Ultraviolet (UV) light has found a pivotal role in HVAC systems, primarily due to its germicidal properties.

1. UV Germicidal Irradiation: By exposing the circulating air to UV light, this method disrupts the DNA of microorganisms, rendering them inactive. The placement of UV lamps is crucial, often near components like the cooling coil or drain pan, ensuring maximum exposure.

2. Photocatalytic Oxidation: A more advanced application of UV technology, photocatalytic oxidation involves the use of a catalyst, typically titanium dioxide. As pollutants come into contact with this catalyst under UV exposure, they undergo a chemical reaction that transforms them into harmless substances.

Integrated Systems for Holistic Air Purification

Modern HVAC systems are increasingly integrating these technologies to provide comprehensive air purification solutions. A combination of advanced filters and UV purification mechanisms ensures that the indoor environment is free from both particulate contaminants and microorganisms.

Furthermore, the integration of smart technology in HVAC systems is enhancing air quality monitoring capabilities. Advanced sensors can now detect changes in pollutant levels, enabling the HVAC system to automatically adjust its operations for optimal purification, reducing the need for manual interventions and ensuring consistent air quality.

As the world grapples with environmental challenges and public health concerns, the role of HVAC systems in ensuring clean indoor air has never been more critical. Advances in filter technology, coupled with the innovative use of UV light, are setting new standards in air purification. As these technologies become more mainstream, individuals and businesses alike can look forward to healthier indoor environments, reinforcing the importance of HVAC not just as a provider of comfort but as a key contributor to well-being.

Advancements in AI and Machine Learning in HVAC Systems

Harnessing the Power of AI for Predictive Maintenance

The age-old saying, "If it ain't broke, don't fix it," no longer holds water in the modern world of HVAC. With the integration of AI, systems have become astute fortune-tellers of their own health, signaling potential issues before they escalate into costly repairs or replacements.

How does it work? Advanced sensors continuously monitor various components within the HVAC system, from compressors to evaporator coils. This data, when channeled through AI algorithms, can pinpoint anomalies, unusual patterns, or performance degradation that often precede a failure. The result? Technicians can intervene early, ensuring minimal downtime and preventing catastrophic damage. Moreover, by addressing issues in their nascent stages, repair costs are often significantly reduced.

Optimizing Energy Consumption through Machine Learning

Energy efficiency remains at the forefront of HVAC system design and functionality, given the ever-growing emphasis on sustainability and cost-effectiveness. Machine learning plays an indispensable role in this context.

AI-driven HVAC systems can assimilate vast amounts of data, from indoor-outdoor temperature variations to occupancy levels, and adjust operations in real-time to ensure optimal energy consumption. Machine learning models can 'learn' from past data and predict future demands. For instance, by analyzing data from a previous heatwave, the system can preemptively adjust cooling parameters to handle a similar event in the future, ensuring consistent comfort levels without significant energy spikes.

User Behavior Analysis: Crafting Personalized Comfort

Gone are the days when users had to continually tinker with thermostats to achieve their desired comfort level. With AI and machine learning in play, HVAC systems have become adept at 'understanding' user preferences and habits.

By continuously analyzing user behavior and feedback, these smart HVAC systems can craft a personalized environment for each user. Let's break this down with an example. Suppose a homeowner usually arrives home at 6 PM and prefers a cooler temperature to relax. The system, having 'learned' this pattern, will start cooling the house shortly before 6 PM, ensuring a comfortable ambiance upon arrival.

Furthermore, these systems can also detect anomalies in user behavior. For instance, if the system recognizes unusually high humidity levels in the house coupled with low activity, it might deduce that the homeowner is away on vacation and optimize its operations to maintain the home's health without excessive energy expenditure.

Synergy for Improved System Longevity and Efficiency

The integration of AI and machine learning in HVAC systems is more than just the sum of its parts. When predictive maintenance pairs with energy optimization and user behavior analysis, the resultant synergy ensures both longevity and heightened efficiency.

For instance, an HVAC system that operates at its optimum, thanks to energy optimization algorithms, experiences less wear and tear. This, combined with predictive maintenance, ensures that the system's lifespan is significantly prolonged. Meanwhile, understanding user behavior ensures that the system doesn't operate excessively or unnecessarily, further enhancing its operational life.

In the labyrinth of ducts, compressors, and coils, lies the brain of a modern HVAC system — AI and machine learning. These technological marvels are gradually reshaping the industry, making systems more intelligent, responsive, and user-centric. By continuously evolving and learning, AI-driven HVAC systems promise a future where comfort, efficiency, and sustainability coalesce into a seamless experience.

YOUR JOURNEY IN HVAC

As we bring this guide to a close, let's take a moment to reflect upon the evolution of your understanding of HVAC. From the initial pages, where you may have felt like a fish out of water, to now, where you stand as a more informed homeowner, your progression is remarkable. This journey you've embarked on, transitioning from a novice to an enlightened individual, is significant and worth celebrating.

Remember, every expert, every seasoned technician, started exactly where you did – at the beginning. Every journey starts with understanding the basics, laying a foundation. Along your path, you've found confidence, not just in grand revelations but in small victories.

Whether it was the first time you identified a hiccup in your system or when you managed to replace a filter successfully, each step was pivotal.

And in moments of uncertainty, you learned the importance of reaching out. The expansive HVAC community, both offline and online, stands as a testament that no one is ever truly alone. Knowledge is ever-evolving, and in the domain of HVAC, the horizons of learning are infinite. Stay curious, stay hungry.

Your newfound comprehension transcends mere knowledge. It's economic empowerment; a means to avoid superfluous expenses and champion energy efficiency. More than that, it's a peace of mind. An intimate understanding of your HVAC system ensures fewer unforeseen breakdowns, guaranteeing comfort.

This journey has been more than just about HVAC; it's been about personal growth, resilience, and mastering a skill that seemed daunting at first. And let's not forget the pivotal role a well-maintained HVAC system plays in ensuring the health of everyone under your roof. Your newfound prowess ensures you stand tall in the face of decisions, be it hiring a technician, investing in a new system, or contemplating upgrades.

But this is not the end. It's a new beginning. A call to keep exploring, keep learning, and perhaps, even sharing the knowledge you've accrued. The realm of HVAC is vast and ever-expanding, with innovations on the horizon and sustainable options becoming more prevalent.

As we conclude, always remember: Your HVAC journey is a reflection of your dedication to your home and your commitment to personal development. Every chapter read, every concept grasped, is a testament to your tenacity. As the future unfurls, embrace forthcoming challenges with the same zest, enthusiasm, and insatiable curiosity you've shown thus far. Here's to every homeowner's continuous journey in the captivating world of HVAC!

Made in United States
North Haven, CT
20 January 2025